No. 1321
$10.95

# ELECTRIC MOTOR TEST & REPAIR

## 3RD EDITION

BY JACK BEATER

TAB BOOKS Inc.
BLUE RIDGE SUMMIT, PA. 17214

THIRD EDITION

SECOND PRINTING



Printed in the United States of America



Library of Congress Cataloging in Publication Data

Beater, Jack.
Electric motor test & repair.

Includes index.
1. Electric motors—Maintenance and repair.
2. Electric motors—Windings. I. Title. II. Title:
Electric motor test and repair.
TK4057.B3 1982 621.46'2'0288 81-18297
ISBN 0-8306-0059-0 AACR2
ISBN 0-8306-1321-8 (pbk.)

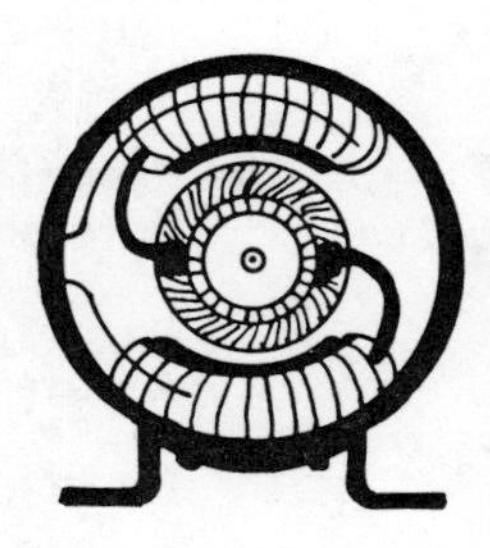

# Contents

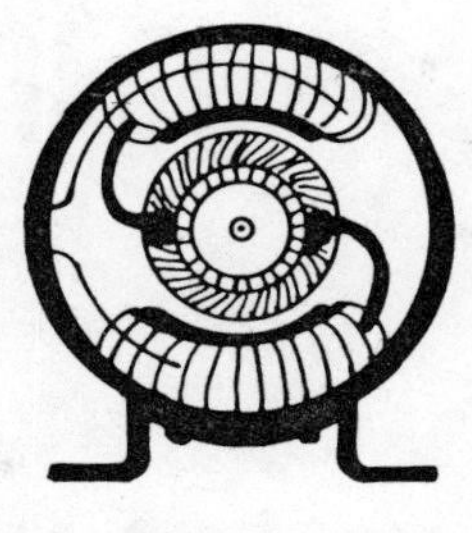

# Introduction

While many of the larger motor repair shops find it more expedient to replace low horsepower units, the rewinding of small electric motors is still a widespread and profitable business. Thus, there is a distinct need for this practical guidebook on the subject.

This is not a college-level, theory-ridden textbook. It is a workshop handbook, written in a style every motor repair technician will appreciate. It contains a wealth of useful information on testing and rewinding small horsepower motors of every type.

# Chapter 1

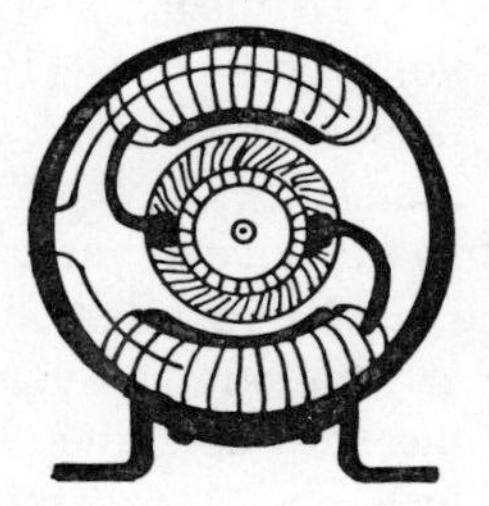

# A Practical Motor Test Panel

When visiting a radio repair shop, one will often find a room cluttered with all sorts of electronic gadgetry such as oscilloscopes, signal generators, and frequency counters along with many other devices used in testing. The automobile service center also has its own array of tools and equipment, designed for the work which is done there. A motor repair shop should likewise be equipped with the necessary devices for testing and checking electric motors.

Electric motors are not tiny electronic circuits, so there is little use for fancy diagnostic equipment. Most of the time, ammeter and voltmeter tests, along with continuity checks and a few other procedures are all that is needed to service all types of induction motors.

A motor test bench is almost a necessity for the shop doing even a moderate amount of motor repair. It has an immense practical value as well as being good advertising. Customers coming into the shop will undoubtedly look around and notice that you use more than a hammer and a pair of pliers to wind a motor or change a bearing.

It takes a good while to get established as a reputable firm in any type of service, and one thing that will help is satisfied customers. Advertising is designed to attract customers, but it is up to the repair man to satisfy them. For this reason, many shops having better than average equipment find that it pays to invite customers back into the shop and spend a few minutes explaining some of the operations and pointing out some of the special equipment necessary to do a good job.

The multimeter is probably the favorite test instrument of all motor rewinders. Multimeters are useful to check continuity and voltage, but that's about all. They cannot be used to measure current because they usually do not have the capacity to handle the amounts of current encountered when working with motors.

To measure voltage, you will be dealing with power line voltages, such as 110, 220, or 440. Thus a multimeter with a great number of voltage ranges is not needed. Finally, on the subject of resistance, most of the time the repairman will be checking for shorts or continuity. When checking for shorts, an extremely sensitive setting is desirable, such as several kilohms or even megohms. When testing for continuity, a very low scale is needed which will measure even the slightest resistance, such as an R×1 scale, or less. Thus a small meter with only one or two resistance scales may not be enough. In any case, the multimeter can be carried from place to place, which is nice, but because of this it should be fairly rugged if it is to last. A permanent, bench test facility is described later.

Most readers of this book will be working with smaller motors (those of less than 10 horsepower). These motors are small enough to be worked on while on the top of a bench, which greatly eases the task of getting to the motor. A motor test bench is essential for the repairing of all types of motors, as it allows motors to be checked in different ways from a central location, without the difficulty of moving the equipment.

A repair shop requires a large, well-equipped bench, while the home service man usually can get by with less space. Figure 1-1 shows the top view and layout of one type of test bench suitable for a small shop. This bench may be of any size, preferably at least 10 feet long and no deeper than 3 feet. The bench should be made of smooth hardwood and very securely braced to keep vibration and noise to a minimum. As an alternative, a layer of padding can be used to help cut down on the noise. A mat of insulating material should be laid on the floor in front of the bench for personal safety.

Depending on what type of tools the shop uses, such as coil winders, there may be other layouts needed. By keeping the number of specific parts to a minimum in Fig. 1-1, the repairman can make his own refinements. Basically, the front of the bench should be left uncluttered so motors can be slid from one end to another without having to be lifted over obstacles. The front side of the bench should have an overhang which allows the use of C clamps to hold motors in place near the edges. A helpful feature might include a ramp that can be placed against the bench, allowing heavier motors to be pulled up onto the bench instead of having to be lifted.

Figure 1-2 shows a motor test panel. This device allows

motors to be hooked into the line and monitored while running, as well as other out-of-circuit checks. This test panel should be mounted vertically at the back of the bench in order to save space on the top. The feed wires can be routed to the top from a remote circuit breaker. A circuit breaker (or fuse box) should be incorporated into the feeder wires somewhere to protect against fire. Circuit breakers, although more expensive than fuses, can be used over and over, thus eliminating the need for replacement each time they trip. Each hot line should be protected according to its voltage and current capacity.

## POWER SOURCES

The line enters the panel and goes directly to two voltmeters, one for the 220 volt circuit and one for the 110 volt circuit. These meters are left permanently connected to the line and give a voltage reading of both single phase lines. They allow the technician to monitor for fluctuations and also to allow specific voltages to be applied outside the boards by means of S3 and S4 via the test leads.

An ammeter is connected in series with the 110 volt line to monitor current flowing through the circuit. This ammeter should be of ample capacity to handle all expected currents encountered in the motors under test. A shunt switch, S1, which is basically a heavy duty spst switch, bypasses the ammeter, when desired, in order to protect it from damage should a large current be drawn through the line. Motors brought in for repairs, or those with possible defects should be tested with the ammeter out of the circuit before taking current measurements.

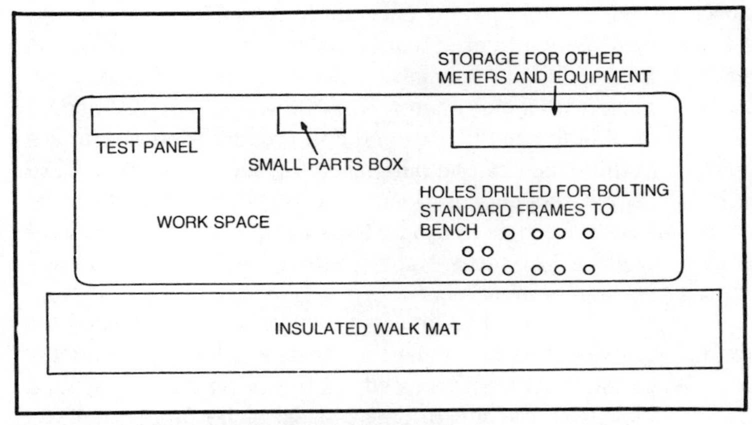


Fig. 1-1. A typical layout for a test bench.

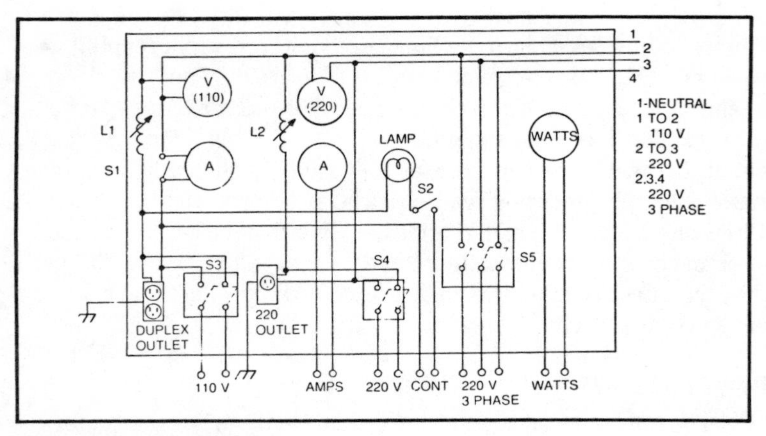


Fig. 1-2. Motor test panel.

The 110 volt line is connected to a grounded duplex receptacle and from there to a pair of test leads through a dpst switch, S3. This allows the hookup of motors without line cords to be made swiftly. The use of a *ground fault interruptor* on the receptacle is highly recommended to guard against shock. Motors with shorts or grounds can be dangerous to touch if the frame becomes the hot side of the 100 V line. Ordinary circuit breakers may not trip but even a small current can prove fatal. The GFIs will sense a leakage to ground and shut off the power.

The 220 volt line is similarly connected to an outlet and a pair of test leads, except that the ammeter is left out of the circuit. This is so the ammeter may be externally connected to either the single phase lines or the three phase line. The meter must be of sufficient capacity to handle the currents encountered in all tests for which it is to be used. No voltmeter test leads are provided outside the bench. If any additional voltage checks must be made, they can easily be made with a multimeter. This allows both on line and off line checks of individual phases and coil groups. Meters that are mounted in the bench can be purchased for between $10 and $20 each, if a little shopping is done. It is important that they be reasonably accurate to allow proper tests to be made. They can be of whatever physical size the operator prefers. The larger meters will naturally be more expensive.

Tapping off of the 110 line is a switch and a lamp used for continuity checks, sort of a substitute or double for the ohmmeter tests. When the switch, S2 is closed, a 110 volt potential is applied to two leads outside the bench. These leads can be clipped across a winding or between a winding and a frame, as in checking for grounds. The lamp should be small (7 watts is fine) to keep the current low in the path so that no dangerous conditions occur.

Variable inductors L1 and L2 are optional. They add considerably to the cost of the panel but are usually well worth the expense. These coils are basically large variable transformers of which only the adjustable side is used. They can be shorted out so they have no effect, or they can be opened up to limit the current flowing in either circuit. Reducing the current in this manner is one of the most advantageous methods of testing motors. It allows even shorted motors to be run intermittently without damage. They are also great for experimenting with motors to compare torque and performance with a less than normal input. They can also be used to prevent the current surge which is drawn as a motor starts, preventing line fluctuations. It is very important to select transformers of sufficient current capacity to allow them to function without overheating. When wide open, that is, with all the inductance in the circuit, they will limit the current considerably. As they are closed, current flow will increase. If a large motor is being tested, and the inductor is closed to the point where the current draw will exceed the maximum allowable in the transformer, then the coil must be shut off completely or it will burn out. Always close the inductors with the power off. See Fig. 1-3.

A wattmeter is also included in the board but is left out of the circuit so it can be connected to any motor under test with clip leads. Figure 1-2 shows only two leads coming from the wattmeter, but some instruments will have four. In either case, the meter is hooked

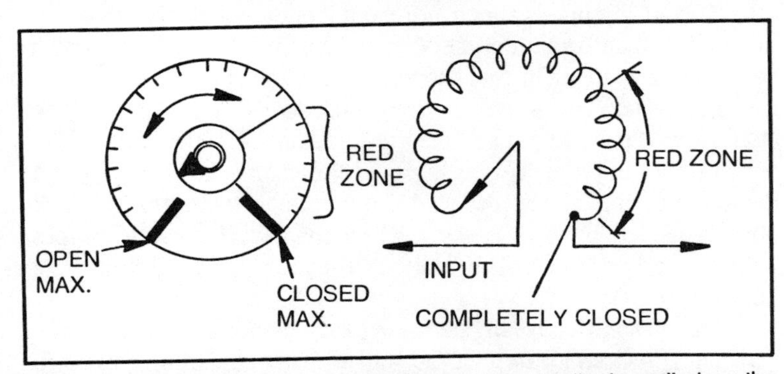


Fig. 1-3. A variable inductor. Never operate the coil in the "red zone" where the inductance is so low that more current can be drawn than the coil can take. To shut off the inductor, disconnect the power and move the pointer through the red zone to the completely closed position.

in series with the motor and also in parallel if it has extra leads. The wattmeter is ideal for measuring peak starting current and voltage, as well as full load power consumption, to allow comparison of the efficiency of various motors.

The 220 volt three phase line is brought out to terminals via a single pole and a double pole switch mounted side by side. The reason a three pole switch is not used is because the arrangement shown makes for easy testing of current flow in any of the three phases. Assuming that knife switches are used here, an ammeter can be connected across the terminals of the single pole switch. The current will remain off until the double pole switch is closed, which will then turn the motor on. To check the current flow in the other two phases, the motor leads may be interchanged at the clips to place any of the three terminals at the single pole switch where the ammeter is connected. Remember that three phase power in this circuit will only flow when two or all three connections are made, so the opening at the single pole switch allows the ammeter to be hooked in without causing any power to flow until the other switch is closed.

All of the switches used on this panel should be heavy duty, as motors can draw considerable current which would damage lighter components. If knife switches are used, it is important that the bare metal tabs of these units do not come in contact with the operators hands, or a severe shock may result. The leads and alligator clips at the base of the panel should be long and heavy, so they can reach the motors on the bench and handle all the current anticipated. The alligator clips should be insulated, of course.

No 440 volt line is shown here, but if one is available, it may be incorporated into another section of the panel using the given design as a guide. On the subject of the voltages, if a motor is capable of operating at two voltages, such as 110 and 220, it may be set for the 220 volt input and then test run on the 110 volt line. The current consumption will be less and the chance of burning out a defective stator will be reduced. Low voltage testing of this and other types will be discussed in another chapter. If rewound motors work properly on the reduced voltage, they can then be tested at the full voltage.

A good addition to this bench in a larger shop would be a large disconnect switch located on the wall which would allow power to be shut off to the entire panel. This prevents unauthorized use of the test equipment and protects anyone who should happen to touch one of the leads or switches while the power was on from harm.

The above equipment makes a very nice arrangement for a bench-mounted test facility to handle all electrical checks. To make mechanical tests, different devices are used. The most common test made in the shop is that which measures torque, such as starting, accelerating, and full load torque. To measure torque one needs some form of dynamometer.

## MEASURING TORQUE

A dynamometer is basically a unit which applies a load to a motor and then measures the torque, or turning effort exerted against the load. Simple arrangements are easy to come by. The most common and the best method for use in the small shop is the prony brake which consists of an arm clamped around the motor's shaft coupled to a scale. As the motor turns, friction of the clamp pushes the arm down on the scale (if a platform scale is used) or pulls on a spring scale. Figure 1-4 shows the construction of the prony brake.

The arm of the prony brake should be made of a stiff material, such as maple or ebony. It should be lined with brake material that will exert a force without a tendency to grab. Naturally, the clamp should be able to grip the motor's shaft firmly. A clamp which fits a very small shaft will not, of course, fit a much larger one. The shaft itself must be round, not flattened or keyed. If it is not round, attach a small diameter pulley to the shaft and then clamp around it. If the arm is heavy, it should be counterbalanced at the short end near the shaft so that it does not add its own weight to the reading on the scale. The scale itself should be accurate and have a range that will take the motor's torque without pinning the pointer. Distance $x$ should be ideally 1 foot, then the readings on the scale would be in foot-pounds. If it is not, the formula to compute torque is as follows:

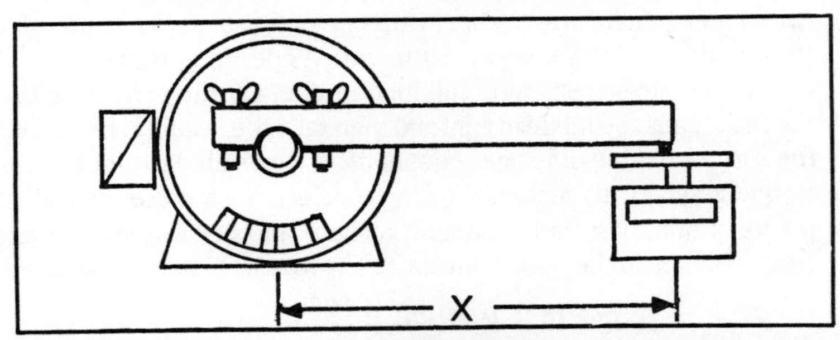


Fig. 1-4. The prony brake. An arm is clamped with thumbscrews to the motor's shaft. Distance X is the distance from the center of the shaft to the arm end.

$$\text{Torque in lb-ft} = \text{scale reading in lbs} \times \frac{\text{distance x (in inches)}}{\text{12 inches}}$$

Before starting to make tests, first be sure of the direction of rotation of the motor, or serious damage may result. The arm must correctly pull or push depending on the scale used. The motor should be firmly clamped or bolted to the bench. Various readings of torque can be taken by tightening or loosening the screws to give the desired load. To measure the locked rotor torque, clamp the screws tight enough so the motor doesn't turn at all and take a reading. It is important to perform this test quickly, as a motor can be locked for only a few seconds before it begins to overheat.

To measure the pull in torque, loosen the screws slowly to allow the motor to accelerate. At the instant the centrifugal switch clicks open, take the reading. This is the point of pull at which the motor can operate up to full speed without the starting circuit being in the circuit. During normal use, if a motor is loaded down too much, the switch may close and place the starting winding back in the line, giving added torque which would then accelerate the motor. Repeated cycles of this sort would burn out the starting winding, as it is not designed to give this type of on/off duty. For this reason, the centrifugal switch is designed to cut out as the motor accelerates to maybe 75 % of full speed, but it clicks back in only when the motor is slowed to about 10 to 20 percent of full speed.

The full load torque must be made at the full load speed, specified on the nameplate. One way to obtain this and any other speed measurement is with a contact tachometer. A tachometer such as this is a relatively inexpensive device with a "tongue" or feeler which is held lightly against a rotating object. A bump on the object (such as a keyway or flat spot on a motor shaft) causes the feeler to vibrate. The device then translates these vibrations into an rpm reading on its scale. Thus, with one of these units it is possible to measure motor speed over the entire operating range.

In the measurement of full load torque, set the screws so that the tach reads the full load rpm and then take the reading. By setting the screws to give other motor speeds, it is possible to plot a speed torque graph of any motor, showing acceleration features as well as points of minimum and maximum torque. Another way to compute torque is by a mathematical formula. The following equation is used:

$$\text{Torque in lb-ft} = \frac{\text{hp} \times 5250}{\text{rpm}}$$

This is a nice way of finding torque if the rpm is known. Of course, if the full load torque is what we're seeking, we just take the full load rpm from the motor nameplate and plug it into the formula, without taking any measurements at all. To use this formula at other speeds we must use the tachometer to measure the rpm, but what about horsepower? Horsepower output varies with speed. For a single phase motor:

$$\text{Horsepower} = \frac{I \times E \times \text{apparent efficiency}}{746}$$

The apparent efficiency is the actual efficiency multiplied by the power factor. Sometimes full load values of these are given on the motor nameplate, but this becomes useless when working at speeds other than the full load. For this reason, the dynamometer test is the most practical.

There are two other methods of measuring torque besides the prony brake. These make use of a cord wrapped around a smooth pulley on the motor's shaft. If one end of the cord is attached to a spring scale, and the other pulled by the hand, a load is applied to the motor and the scale reading can be taken. See Fig. 1-5 for details of this method. As an alternative for the hand and the spring scale, a weight can be used instead. In Fig. 1-5C, a bucket of known weight is attached to a cord which is wrapped several times around a smooth, flat pulley. By filling the bucket with various amounts of sand, the weight can be adjusted to give the desired load to measure either the starting, the full load, or other torques along the acceleration curve. The bucket is then removed and weighed to give the reading.

The rope used should be small in diameter and stretch resistant. Woven linen is a good choice. The pulley and the rope should be waxed so the two do not grab or the measurements will be very difficult to take. The diameter of the pulley is not critical, but its diameter must be accurately known. Just as with the prony brake, for which the length of the arm is a factor in the computing of torque, the diameter of the pulley and the thickness of the line is the equivalent mechanism. See Fig. 1-6. The distance $x$ in Fig. 1-6 is equal to one half the pulley diameter plus one half the rope diameter. If the measurements are in inches, they must be divided by 12 to convert them into feet (so the end result will be in lb-ft).

To test a ¼ hp motor, 1725 rpm, using a 4-inch pulley and a ⅛ inch diameter rope, use the following formula:

$$\text{Brake arm (distance x)} = \frac{\text{pulley dia. in inches} + \text{rope dia. in inches}}{12 \times 2}$$

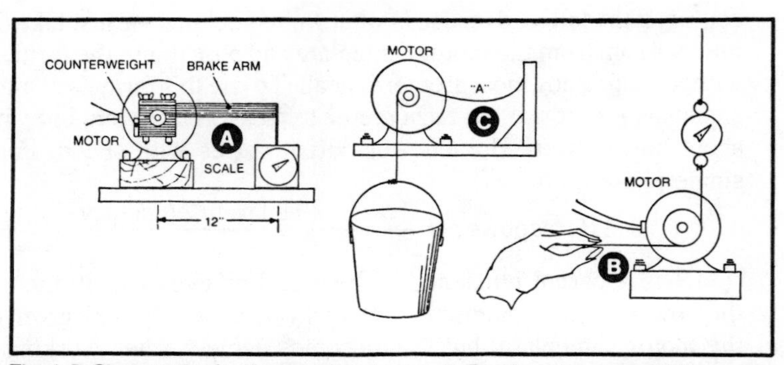


**Fig. 1-5. Shop methods of measuring torque. In *B* pull on the rope until the motor turns slowly, read scale, and compute starting torque according to rule. In *C* add sand to bucket until motor turns slowly and takes up slack in rope. Weigh bucket and then calculate torque according to formula. In *A* the prony brake is shown.**

$$= \frac{4 + 0.125}{24} \text{ ft}$$

$$\text{Starting torque in ft-lb} = \text{brake arm} \times \text{wt. hung on rope}$$

$$= \frac{4.125 \times \text{wt.}}{24}$$

$$\text{Full load torque} = \frac{\text{full load hp} \times 5250}{\text{full load rpm}}$$

$$= \frac{0.25 \times 5250}{1725} = 0.76 \text{ ft-lb.}$$

Notice how small this full load torque is for the ¼ horsepower motor. Is that surprising considering torque wrenches are often graduated up to 120 lb-ft? Not really. The rpm is what determines the torque at a given horsepower output. If the motor were geared down to only a few revolutions per minute, the torque would soar, giving a more significant amount. Try experimenting by putting different values in the formula for the speed and see how they affect the outcome. These speed changes would have to be made by gearing or belting, because actually changing the motor speed by changing the load would change the horsepower, so the formula would not work. This exercise helps to reinforce the idea of the actual power coming out of the motor, which is a significant amount.

Recently there has been a lot of work going on to develop quiet motors. Noise levels produced by some motors must be kept to a minimum for environmental reasons, in certain uses, so the sound pressure level of the motor is a useful bit of information. Noise is generated in several different ways. One is directly from the bearings. Sleeve bearings are inherently quieter than ball bearings, and ball bearings can easily become marred which produces still more noise. Another source of noise comes from rotors which are out of balance. Usually, rotors are balanced at factories when motors are assembled but occasionally it is necessary to drill out a small amount of material on one side of the rotor to eliminate a defect. Sound is also produced by armature pulsation and by the primary laminations vibrating in the magnetic field produced by the winding. Single phase motors are more prone to this than three phase motors because in the former, there is more "jumping" action of the flux, while in the latter, the flux rotates continuously. Air rushing inside a motor can also be a source of noise, but this is usually minimal. Also, cooling fans set up vibrations which are quite muffled compared to that of the motor vibrating due to the pulsating flux.

Some motors are inherently noisier than others. Frame size as well as internal construction gives each motor its own particular ability to resonate at a certain frequency. While it could be said that larger motors dampen vibrations more effectively than smaller ones, this is not always the case. Sometimes a motor with perfect bearings will echo sounds that make it seem like there is something wrong when in fact, there is no fault.

A sound level meter is a useful tool to measure the noise give off by different motors. These are calibrated in decibels, which can be read directly on a built in scale. Sometimes it will be necessary to obtain a particularly quiet motor, while at other times almost any one will do. A sound meter is therefore desirable to make comparisons between motors. Like the multimeter, they are small enough to transport around.

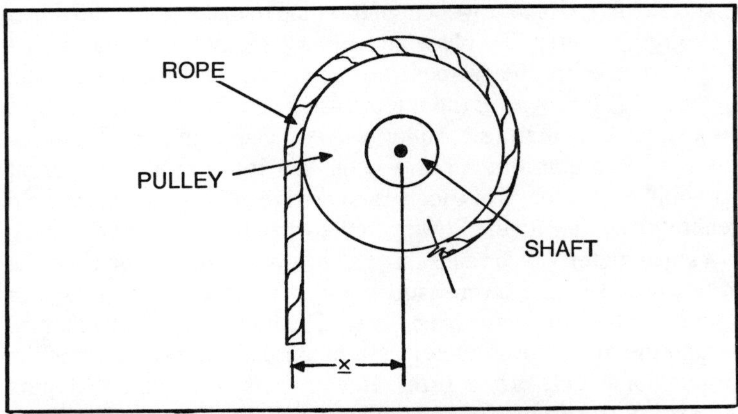

Fig. 1-6. Distance X is the same type of measurement as in Fig. 1-4.

The tools mentioned throughout this chapter are adequate for nearly all small shops and home applications. For lab or experimental work, other equipment is needed. A lot is up to the individual, depending on what his needs are and how much money he can spend. The test panel can be elaborated on, the bench can be redesigned, other tools such as lifts and cranes can be added to the shop to facilitate handling of larger motors, or the entire version can be scaled down. One tool not mentioned previously as a "test" instrument is a grease gun. Another is a bearing or pulley puller. These have a definite place in any shop. Other machine tools such as grinders, lathes, drills, etc., should be on hand in a larger shop, as their use may mean the difference between a day's work and an hour's.

A large, variable transformer and a variable inductor like the one mentioned earlier are some of the most helpful devices to any motor repairman, as their use in the diagnosis of burned out and rewound stators makes for easy testing without the fear of damaged motors. Old motors with burned out stators can be rewound for the sole purpose of being a current limiter (an inductor, but fixed) for the testing of motor windings. The choices are up to the serviceman.

Safety with any device is of prime importance. With all electrical instruments, there should be no exposed parts carrying a high potential. Mechanical tools should be in good working order to prevent failure. Motors under test should be securely fastened to the bench (particularly when making torque tests) to prevent them from jumping around and hurting the operator.

There is one more item which is useful to a motor rewinder, especially if he has to travel out of the shop to make tests, and this is a clip on ammeter. To place an ordinary ammeter in line with a motor, one of the leads must be opened up and the instrument connected. Moreover, if the motors are large, the current will also be large, necessitating a bulky and expensive meter. A clip on ammeter is a unit which has a hook that loops around a current carrying conductor. This loop is sensitive to the magnetic field generated by the current flowing in the wire. It then sends a reading to a small meter which reads directly in amps. Thus, large currents in heavy wires can be measured without the need of cutting or disturbing the conductor in any way. The meter is simply clamped around the flow. These meters can be purchased from most of the larger tool and electronic firms. Buy one with the range that suits the size of work most often handled.

# Chapter 2

## Tools for the Motor Rewinding Shop

As in any other trade the right tools play an important part in the success of a rewinder or motor repair man. Men long in the business have developed special tools for certain operations and have acquired others by purchase. To turn out work efficiently—and at a profit—requires the use of labor-saving equipment and production methods. The need for special tools in the small motor shop is as great as it is in the factory, although the kind of tools needed for the motor shop will be simpler and less expensive than those needed in the process of manufacture.

We have all known mechanics who seem to take pride in the fact that they can get by with only a screwdriver, hammer and pliers. And "get by" is about all they can do. Thomas Edison is once said to have remarked, "It is surprising to see what a good mechanic can do without tools, it is surprising to see what a poor mechanic can turn out with the aid of good tools, but when you get a combination of a good mechanic and good tools—well, there's no limit to what he can do."

The winder or motor repair man in the small shop need not go without the necessary tools to do good work and speed up his output. Many of the most necessary tools can be built right in the shop in spare time, and the others can be purchased at reasonable cost. Many a man starting in the business with little capital has built up a good part of his tool equipment himself, and with the money earned from it has gone out and bought more and better equipment as his business grew.

The purpose of this chapter is to acquaint beginners in the winding and motor repair business with a few of the tools that are almost indispensable for the small shop. These are in addition to other equipment, such as testing outfits, armature and stator holders, and winding benches. Many of the items listed here can be bought from jobbers or manufacturers of shop equipment, such a course being recommended where possible. The illustrations are typical of the kind of special tools that have been found practical for the shop specializing in the repair of the smaller types of motors.

Probably every winder at some time or other, has thought about the use of an armature winding machine. Large electrically operated armature winding machines are on the market and are being used with success in many a shop. They are practical and well worth their cost where there is sufficient volume of business to justify their use. Such machines, with a capacity of over 100 armatures per day, would be out of place in the small shop where the daily average of rewind jobs might be as low as two or three.

On the other hand, a simple hand-operated winding machine is a very useful fixture for the small shop, and also in the large shop for special work. One great advantage of the hand winding machine over plain hand winding lies in speed. Another is the fact that, with a suitable tension device, the coils can be wound with an even tension and the turns accurately counted. The latter is particularly important where the coils consist of a very large number of turns.

## HAND WINDER

Figure 2-1 shows a diagram of a shop-constructed hand winder. The base can be made from some sort of pipe, a large connecting rod from a discarded motor, or from wood. The shaft and handle can easily be made or salvaged from junk. The face plate should be ½ inch by 2 inches, either iron or steel, and about 18 inches long. From each end a slot runs almost to the center so that the armature shaft holding brackets can be adjusted to fit armatures of all lengths. The armature to be wound is held in notches and fastened with U clamps. By turning the machine around it can be made to suit right or left handed operators. The turn counter registers each revolution of the armature.

As indicated in the upper left hand corner of Fig. 2-1, strip insulation is best for use on this kind of winding. Individual slot insulation can be used but is likely to cause the operator a great deal of trouble, as the wire has a tendency to catch on the corners of the paper. Strip insulation has another advantage in that it tends, because of the support from adjoining slots, to remain in place while the slot is being filled instead of sliding out at one end.

In operation the wire is first looped over the star collar on the upper end of the armature shaft, then fed into the first slot, around the drive end of the shaft to the other side of its span, and so on until the full quota of turns has been placed. The end of the wire is now looped over another notch in the star collar and the second coil is wound. This is repeated until all coils have been wound. At the completion of the winding it will be found that all commutator leads are over the star collar under tension and of ample length. The wedging should be done before cutting the leads. Bringing out the beginning ends of the lower coils on top can be done by cutting the wire and placing as the winding progresses.

As the wire is wound into the slots the tension is regulated entirely automatically, the free hand being used merely as a guide to see that the wire enters the right slot. In making a tension device care should be used to see that the enamel covering of the wire is not cracked by bending too sharply. If a device similar to the one shown in Fig. 2-1 is used, the grooved wheels over which the wire passes should not be of too small diameter. Some rewinders use a brake shoe with spring adjustment that applies pressure directly on the rim of the reel. The object of any device of this type is to maintain a constant and even tension at all times, and to prevent backlash.

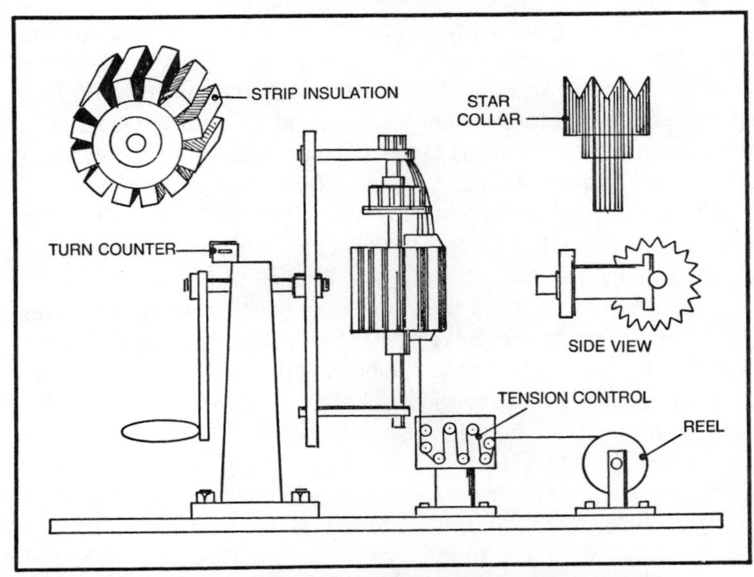


Fig. 2-1. Detail of a simple hand-operated armature winding machine.

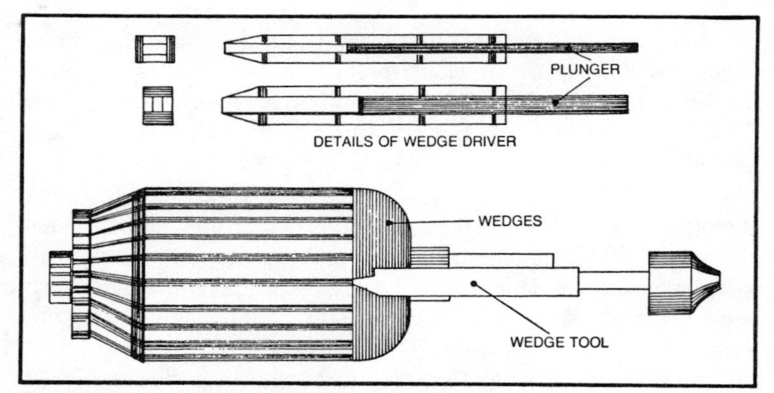

**Fig. 2-2. Wedge driving tool for pegging armature slots.**

When winding armatures with more than one wire in hand, additional tension devices will be needed, one for each reel. The tension should be exactly the same on all wires, otherwise the wire with the greater tension will tend to pinch and bind the sides of the looser coil. When winding with more than one wire in hand, the additional wires should have a tracer so that the proper coil ends can be located without trouble. After some practice the rewinder, with the aid of the hand winding machine, should be able to double his output over the straight hand winding method.

Another great aid to the winder is in the use of a wedge driving tool. Many of the smaller, and some of the larger, cores were designed for fibre wedges. Fibre wedges, being flat, take up less space than do the usual maple wedges, and as a consequence it is sometimes difficult to insert the wooden wedges over a full winding in the space allowed. It is also true that fibre wedges are brittle and tend to buckle and break when driven without the aid of a tool. In very tight spaces the wedge tool is a big help in driving home wooden wedges also.

Figure 2-2 shows the details of a serviceable wedge driver in cross-section and in use. Several sizes will be needed to accommodate the various wedge sizes. The body of the tool is made of brass or steel strips riveted together. The sides of the tool should be slightly thicker than the thickness of the wedges to be used, while the top and bottom of the tool can be cut from ⅛ inch stock. Small steel pins, about 1/16 inch in diameter, are used to rivet the parts together. After assembly the working end is ground down thin so that it can be held close to the core without cutting or chafing the winding. The plunger or driver should have a free running fit in the slot and about ½ inch longer than the body of the tool. In using, a wedge is inserted in the slot of the tool and the tapered end of the driver is held close to the slot entrance. A few light hammer blows on the plunger head will force the wedge into the slot without spreading, breaking or buckling.

Another great time saver for the rewinder is a cutting and gauging board for preparing the cell insulation for the slots. A board of the type shown in Fig. 2-3 enables the winder to easily and accurately trim up a set of insulation to the desired size in a few moments' time. The adjustable gauge at the cutting end, when once set to the right length, allows the operator to do the cutting at high speed without error. The scale, marked off in inches across the top of the board, makes for quick measurements.

The foregoing cutting board can readily be adapted as a machine for forming slot cell insulation to shape. To do this a sheet of thin metal is folded once around a section of the cutting knife, leaving the edge slightly rounded instead of sharp. The slot paper is creased between this dulled portion of the blade and the edge of the board. Only one crease can be made at a time, but it can be made accurately and quickly.

Every winder knows how hard it is to pack down the wires in slots having very narrow openings so that the full quota of turns can be put in place. On slots having wide openings a square of fibre board can be used to press down the wires, but on many armatures and stators the slot openings are too narrow to make good use of such a method. To overcome this difficulty it will be worth the

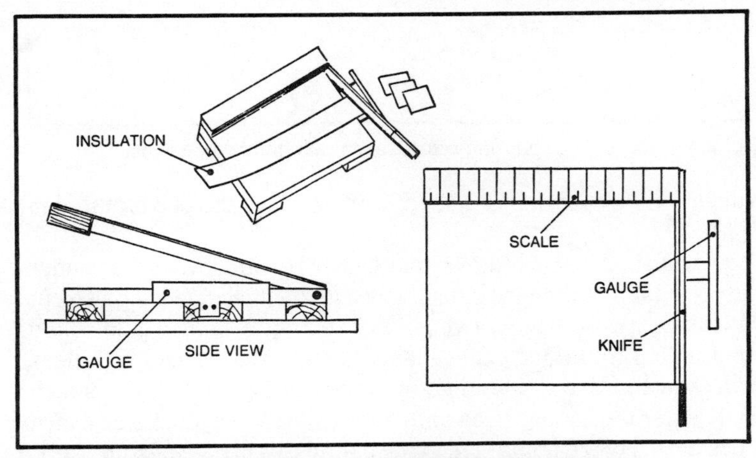


Fig. 2-3. Cutting and gauging board for trimming slot insulation.

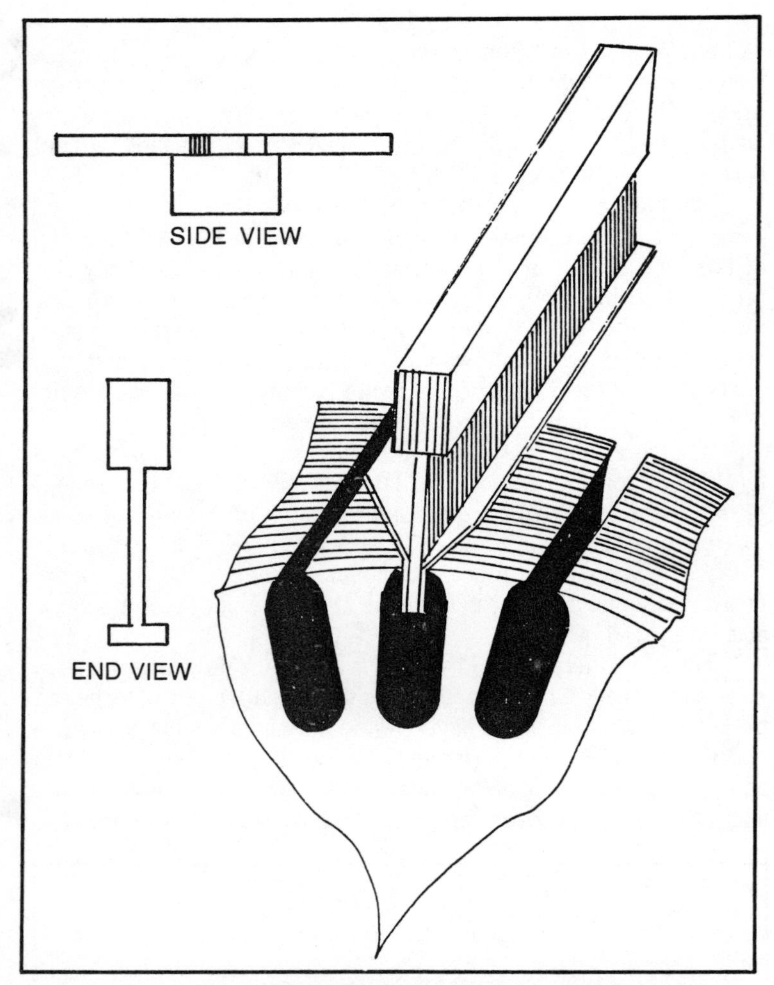


Fig. 2-4. Metal drift for packing coils in slots with narrow openings.

winder's time to make up a set of drifts such as the one pictured in Fig. 2-4.

Drifts in the shape shown may be hard to obtain, but the winder can have them milled from a solid bar at any machine shop. To cut down on expense have a twelve-inch length of ½ inch cold rolled steel milled to the indicated shape, leaving the shank a little less than ⅛ inch. Have the footing, or working edge, left about ⅛ inch thick. After the bar has been milled it can be sawed into three 4 inch pieces, each piece being riveted or bolted to a handle of some sort.

The handle will make it possible to use the drift inside small stators and a good job can be done without the use of hammer or mallet.

After the three drifts have been supplied with suitable handles the foot, or working end, can be ground or filed down to any desired width. For work on small motors, ¼ inch, 5/16 inch and ⅜ inch will be the most useful widths. By means of a small, fine file the edges and corners should be slightly rounded and smoothed so that there will be no danger of damage to insulation of wires or slot paper. A set of these drifts will save a lot of time and make for a neater and more compact winding.

Covering armature coils and field coils with insulating tape is a job that takes considerable time if done with the bare hands. In the larger shops where there is a considerable volume of this work a motor driven coil taping machine is often used. Operating on the same principle as the machines which wind the paper wrappings on automobile tires, a coil taping machine can make a lot of money for the shop if volume permits. However, there are few shops that have sufficient work of this kind to justify the outlay for a taping machine.

Figure 2-5 shows a device for use in the hand taping of motor and generator coils. With this tool the taping job is speeded up considerably as there is no time lost in drawing through a long length of tape each time a turn is made, or else in chasing over the floor after a runaway roll that has escaped the winder's hands. A tape winder such as described will not work with coils having a

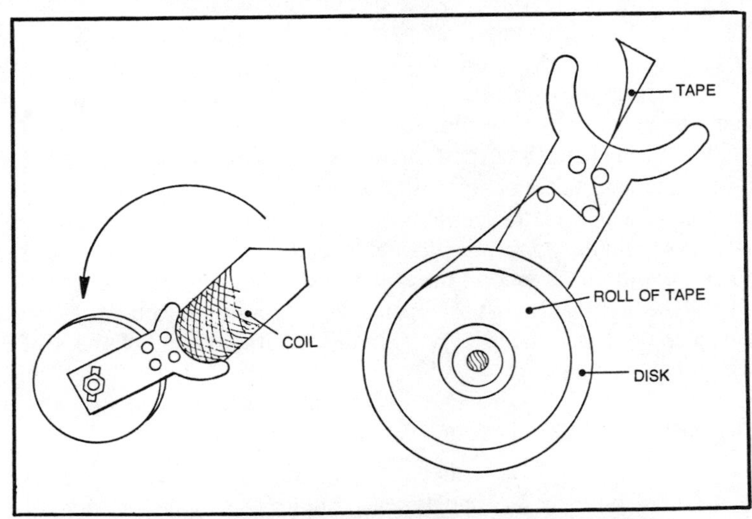


Fig. 2-5. A device to aid in hand taping of armature and field coils.

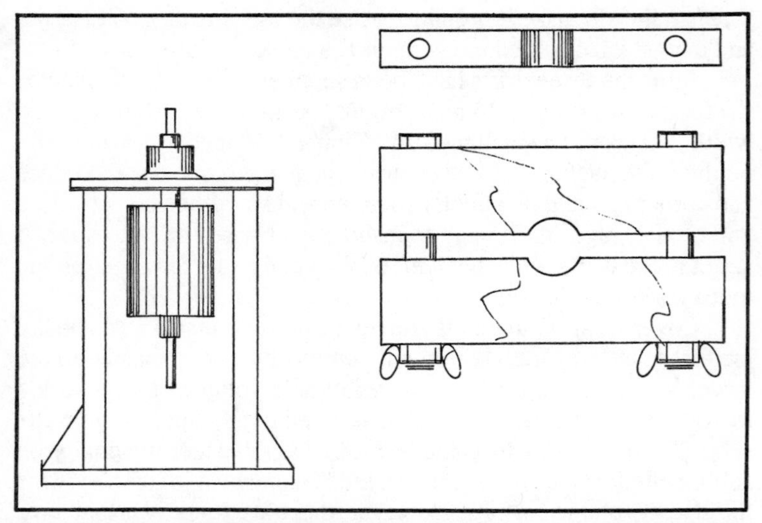

Fig. 2-6. Puller plate for removing commutators and, at the left, a stand for using the puller in a press.

center opening of less than about 4 inches, for the obvious reason that the tape winder will not go through the hole.

Essentially, the tape winder is a means of holding a roll of tape under the proper tension, so that as the winder is rotated around a side of the coil it distributes the tape both evenly and firmly. After the first turn or two the operator has only to flip the winder around the coil, handling it only once per revolution, and seeing that as it progresses the right overlap is made with each turn.

In making a winder the two side pieces can be cut from thin sheet brass and sawed or filed to shape. Four brass machine screws hold the two sides the proper distance apart and also act as guides for the tape. The rool of tape is centered on a through bolt which, by the use of a wing nut, acts as a tension regulator. Two thin fibre washers, slightly larger in diameter than a full roll of tape, are placed one on each side of the roll for support. The overall length should be kept as short as is possible for the reason that this measurement is what determines the minimum size of the coil which can be taped.

## PULLERS

Another group of tools needed by the motor repairman consists of a set of pullers. No one type of puller is satisfactory for all purposes. Roughly, the puller needs of the average small motor shop can be classed as follows: Puller for small flat and V pulleys, puller for large flat leather, steel or composition pulleys, puller for small bronze bushings and a puller for commutators.

Figure 2-6 shows a puller plate for removing commutators from armature shafts. The plate is in two parts, held together by two wing bolts, and notched in the center to fit closely to armature shafts. This plate can be used in conjunction with several types of pullers, and finds its best use in its ability to pull commutators without subjecting the bars and insulation to strains. The plate should contact the inner commutator sleeve and apply the pressure at that point where it joins the shaft. On the left side of the figure is shown a wooden stand to be used with the puller plate when the commutator is to be pressed off in a screw press, or when it is necessary to remove the commutator by hammer blows. The same plate can be used to press commutators on the shaft.

Figure 2-7 shows two common types of pulley and gear pullers. The one on the left is best suited to the smaller types of pulleys and gears, and is well adapted to removing V pulleys similar to those used on most refrigeration motors. Pullers of this type can be purchased which are self locking when under pressure. Note that the puller jaws have projections which extend under similar projections from the sides of the central nut. It can readily be seen that as soon as the screw presses against the nut, the nut in turn presses against the jaw projections, thus forcing the jaws in toward the work

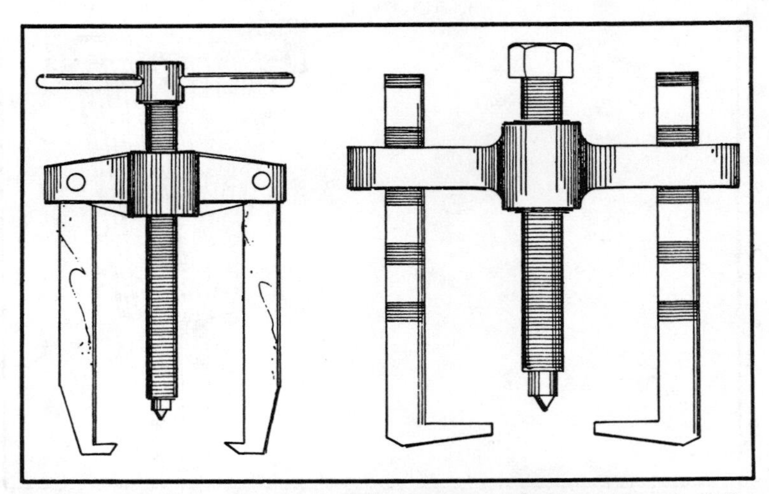

Fig. 2-7. At the left is a puller for gears, V pulleys and other small parts. At the right is a puller for large leather, composition and steel pulleys.

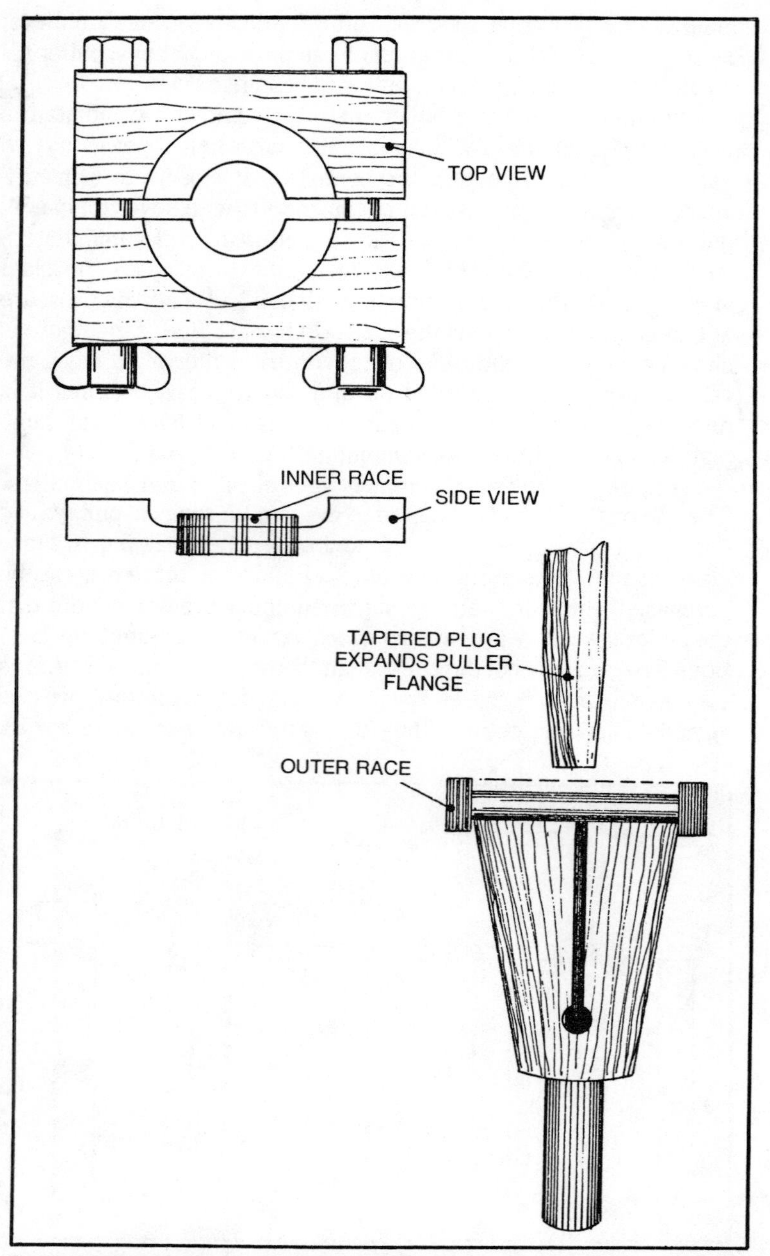


Fig. 2-8. At the left is a puller plate for removing inner ball races from shafts, while at the right is a tool for removing outer ball races from recessed holders.

they are holding. The greater the force needed to move pulley or gear the greater is the holding ability of the jaws.

The puller illustrated at the right in Fig. 2-7 is particularly useful on the larger flat pulleys used on multi-horsepower motors. The jaws are adjustable for width by sliding in the slot of the puller head, and they are adjustable for length by making use of any of the notches provided in the sides of the jaws. The lips of the jaws should be long enough to reach over the leather or composition surfaces of built-up pulleys and to reach the metal hub. The jaws on this type of puller can be reversed, making the tool available for inside work, such as pulling rings and working in recesses.

Figure 2-8 gives some details of a common type of puller plate used in removing ball bearings of the three-piece type (magneto type). The plate, similar to that shown in Fig. 2-6, has a thin lip around the center opening that will clamp tight in the ball groove of the inner race. This plate can be used in a press or with a puller of conventional type. A tool for pulling outer ball races is also shown. The construction of this outer race puller is more complicated than is that of the inner race puller. While such a tool could be turned out by a good lathe man, it would probably be cheaper in the long run to purchase such a tool from the service tool list of some manufacturer. A tool of this kind will only fit one size of bearing, but adapters for larger sizes can be turned out and used in connection with it.

There are many other tools that are equally handy for the rewinder and motor repair man. Many of these are made from old hacksaw blades, wood, fibre, bits of brass or steel, and fashioned to suit the user and to fit some particular job or operation. In fact, the good workman who takes pride in his job is always making something, and sometimes the result of an hour's experimentation will result in a labor-saving gadget that, in the course of time, will return big dividends for the time and effort spent on it.

# Chapter 3

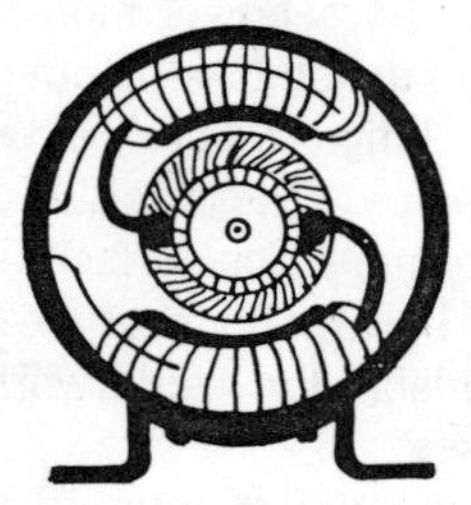

## A Time Saving Coil Winding Machine

Motor rewinding is currently being handled in two main divisions which originally started out as one main operation. Primary burn-outs are the most common form of motor damage (far more common than, say, locked bearings for example), and thus is the main reason for motor repair and rewind shops. Today there are large shops which service the need of the industrial field, rewinding large poly-phase motors and generators. There are also small shops that repair motors of mostly five horsepower and less, the majority of them single phase. Most of these small, single phase motors come from private homes. Furnaces, refrigerators, and appliances are always feeding the business with burned out motors at one time or another. Large organizations that specialize in industrial motors have little interest in fractional horsepower motors because of the small sum of dollars involved in such work. In fact, many larger shops refuse to accept single phase motors for repair.

Small shops, which range from nothing more than a basement in a neighbor's house to a small business located up-front, often do just the opposite. Small shops do not have the heavy equipment to handle large motors nor the equipment to repair them. They are however, ideally suited to work on small motors. Even someone who does not repair motors for profit can generally rewind motors around the home with a little practice. Since several million fractional horsepower motors are being built each year, there is a great business potential in motor rewinding.

The profit that can actually be made is dependent on competition from other rewind shops and the price of new motors. Motor service men must charge for labor on a competitive basis if they all wish to stay in business. Also, for each motor entering a shop for repair, is it actually cheaper to repair the motor or sell the customer a new one? These factors affect the amount of profit that a repair shop can make.

At this time, a ¼ horsepower split phase motor may be purchased for around $40, give or take some depending on the exact motor. Repair shops will often make it a rule never to allow the repair cost exceed two-thirds of the price of a new motor.

The key to low cost repair of motors is to turn labor into an automated task to save time and yet keep the cost of overhead low. The sooner the shop gets in the black, the lower the prices will drop. At a given price of repair, the trick then becomes to get the greatest number of customers in the area as possible. Fast, quality service is one way to do this. Experience is the real way in which this becomes a reality.

Since spending work time to rewind a motor before a certain deadline increases the cost of the job, it is sometimes recommended that the small motor shop keep a supply of rebuilt motors on hand at all times for exchange. Then, when a customer wants a motor fixed in a hurry, he can simply purchase a rewound motor over the shelf instead of waiting for his to be repaired. The service man then has more time to rewind the customer's old motor during the times when business is slow. In this way he can keep the business going without becoming necessarily bogged down with work. Instead of waiting around all day for a repair to come in and do it then, he can exchange the burned out motor for a rebuilt one on hand and then rewind the old motor when he has the time, putting it away as a rebuilt motor for future trade-ins.

If a motor is completely dead, ie., in need of new primary, a new rotor winding, a new set of brushes (the latter two in commutator or phase wound motors) bearings, and so on, it is always recommended to sell another motor as a replacement rather than to rebuild the old motor for the same customer. Quick and dependable service is one certain way of retaining good customer relations, and this can be accomplished best by maintaining an assorted stock of new and reconditioned motors.

Up to now we have been discussing motor rewinding as a business, but what about motor rewinding for one's self? The principles of advertising and customer relations do not apply, of course, but the need for experience and labor saving devices are still present.

For the assistance of those who are new to motor rewinding, we will describe the stator windings of the more common types of

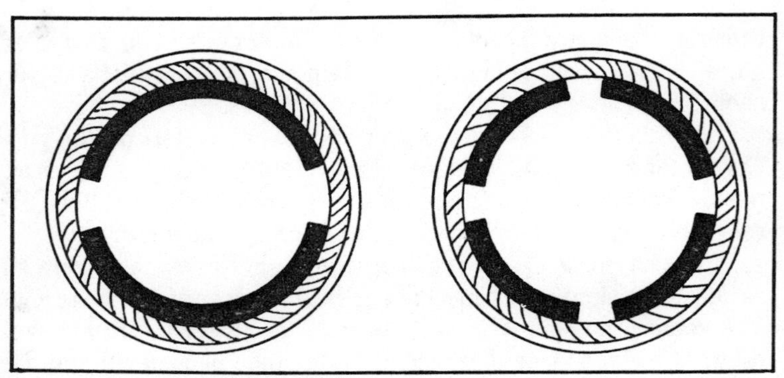

Fig. 3-1. Two and four pole stators.

single phase motors: The repulsion start, induction run, the split phase, and the capacitor motors. In the former there is but one winding on the stator, while the other two contain two field windings, one for running and one for starting. Burned out starting windings are the most common problems with the split phase motor, while in the capacitor motor, either may be defective.

In motors of these types, the section of the stator laminations surrounded by one group of coils becomes a pole. For example, if the total winding is wound in the form of two groups, the motor becomes a two pole machine. If wound in four groups, the motor becomes a four pole machine, and so on. Figure 3-1 shows the end views of both a two pole and a four pole stator. A pole, is shown in Fig. 3-2. These poles may be of one, two, three, or more concentric coil forms, three being shown in the illustration. The coils may be wound from the center out or vise versa, with varying number of turns in each coil depending on the exact motor. These coils are held in place inside the laminations by paper or fiber wedges are pressed into the slots over the wire.

If the stator were rolled out flat so that one could see all the poles in a plane, it would look similar to Fig. 3-3. Each consecutive pole is wound in the opposite direction from its adjacent pole, that is, one clockwise, the second counter clockwise, and so on. Note that there is always an even number of poles, so that opposite pole *pairs* are always formed.

The pole groups are wound with a single, continuous length of wire, starting at the inside of one coil. When winding by hand, the proper number of turns are wound on the slots to be occupied by the inner coil, then the wire is carried into the adjacent slots on each side and the second coil is wound. This is repeated until the full number of coils required in the group is completed. The wire is then brought to the location of the adjacent pole and it is wound in the same fashion. After all the poles are wound, the wire is then brought outside the coil groups and connected to the line terminals. The polarity of each coil is determined by the direction in which it is wound. As mentioned before, the poles are wound in opposite directions from each other, forming discrete poles. Rarely, two adjacent coils may be wound in the same direction, and the two coil groups combine their polarity to form one pole. In Fig. 3-3, winding the coils in opposite directions would form four poles, as illustrated. Were adjacent groups wound in the same direction (two cw, the other two ccw), the result would be a two pole winding.

## WINDING STATORS

In many cases, the hand winding of stators is a very tedious and time consuming task, because of the lack of room to maneuver inside the core. This is because of the sharp edges of the slots and laminations (which have a maddening tendency to catch and strip finer gauge wires) the "spring" of the wire, the problem of kinks, and the difficulty of keeping the proper tension while keeping the wire inside the slots. This is particularly true of two pole windings, where the poles often span 160°. Pulling the wire in one slot yanks it out of the slot on the other side of the coil.

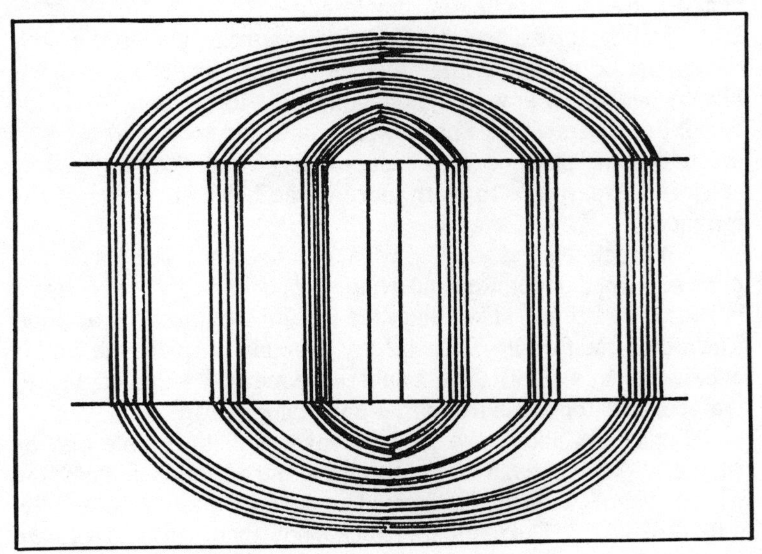

Fig. 3-2. A group of primary coils.

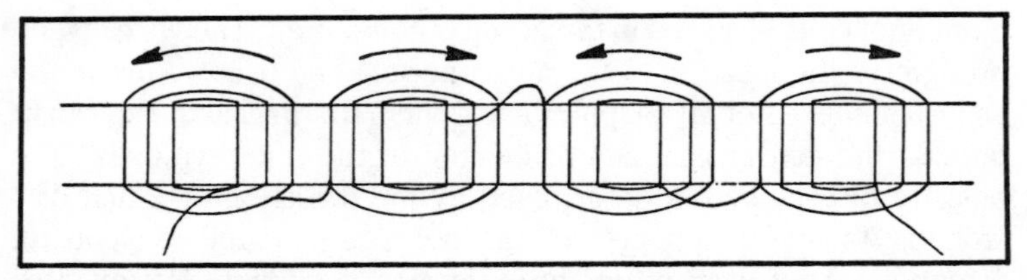

Fig. 3-3. A rolled out stator would resemble this. This particular one is a four pole winding.

The use of a winding gun or bobbin is a great improvement over the straight hand winding, both in time saving and in the neatness of the work. Using a guide like these to spool the wire into the slots allows more control and freedom to move around inside the core. Only long experience and practice enables a winder to turn out a neat, symmetrical job by the plain hand method. Figure 3-2 shows the general appearance of a coil group wound by either the hand or the gun method. Note how the coils rest upon each other outside the slots. This increases the chances for shorts and grounds, should the insulation become damaged. Some rewinders make use of curved blocks between the coils during the winding to space the coils properly. Also, the use of a spacer between the laminations and the innermost coil reduces the chances of the coil "tightening up" and raising out into the path of the rotor. The blocks are removed, of course, after the winding is completed.

Most factories use machines which make preformed coils which can be placed in the slots for a near perfect fit, and this chapter will show how to build a simple device at home to do the same. For those who are new to the game (and some of those who are not) some practice with hand winding would be an excellent source of experience to learn more of the hows and whys of coil formation.

To practice hand winding coils, one need not have an assortment of burned out motors and a huge supply of magnet wire. If this is your first attempt at winding, the results will probably be poor. The motor itself would not be ideally rewound and since the wire is expensive, so we will show a substitute method which duplicates the actual stator and wire with a minimal expense.

Figure 3-4 shows the form in which one stator pole may be wound, showing the number of turns and the direction in which the wire is wound. This winding would be described as a 28-34-42-30-0-0-30-42-34-28. The numbers indicate how many turns of wire are wound into each slot. The two zeroes at the middle indicate that two slots are empty. Not that there are always pairs of identical numbers; two 30s, two 42s, etc.

To make a practice stator, obtain a block of wood perhaps 1 inch high, 2 inches deep, and 12 inches long, as shown in Fig. 3-5. Using a thick bladed saw, cut a series of slots in the wood about ⅜ inch deep, spaced ⅜ inch apart. Using a ¼ inch drill bit, drill holes through the wood at the base of the slots so that finished product resembles a rolled out stator. Sand the block smooth after the cutting and drilling is done. This block is not only a good way to practice, but also a fine method to plan coils and spans when rewinding polyphase motors.

Monofilament fishing line is an ideal substitute for magnet wire when working with this block, being very similiar in feel to the wire and considerably cheaper. Something in the 15-20 lb test range works nicely. For a start, try your hand at winding the coil shown in Fig. 3-4, without the aid of any tools. Do as neat a job as possible. It doesn't matter where on the block you work, near the ends, the center, or wherever. Avoid kinks in the line like the ones shown in Fig. 3-6. They are a definite trouble spot!

After completing this group, unwind the coils back on to the spool and start again, this time using small pieces or corrugated cardboard as spacers between each coil. In a real winding, the spacers would be used to keep the ends outside the slots from

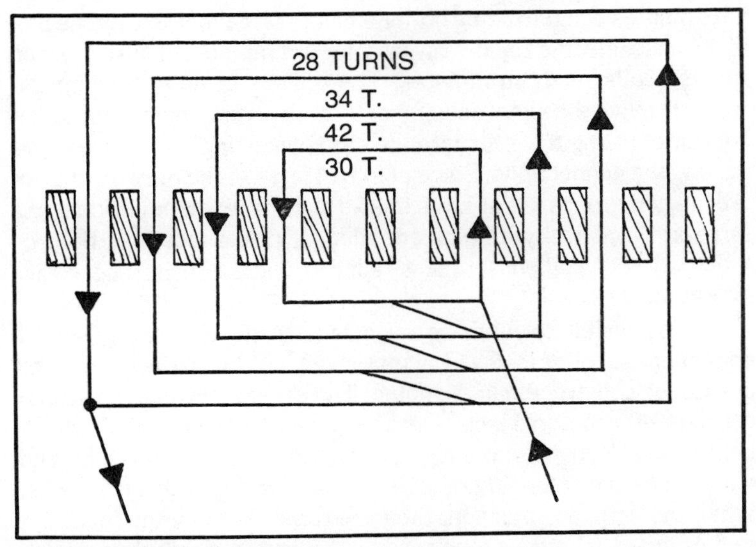

Fig. 3-4. A typical coil group.

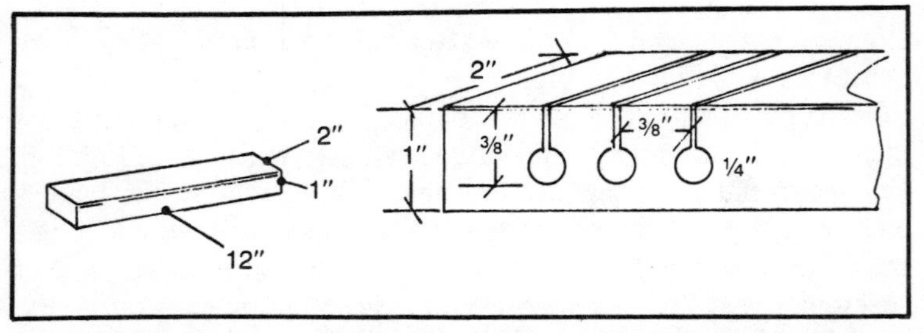


Fig. 3-5. A block of wood, prepared as shown, makes an excellent practice stator.

touching each other. When the pole is finished, the spacers may be removed and the coils adjusted into place, if necessary. Spend as much time as you want winding other patterns of coils on the block, to get the idea of how the job can be done by hand. When you feel qualified (or exhausted) you are ready to take the next step in the winding of motors.

## PREFORMED COILS

Most factories, and many rewinders, use a method of rewinding that has many advantages over the straight hand method. This consists of the use of preformed coils prepared by means of specially designed winding machines. Such a winding machine is one of the most valuable tools a motor repairman can have, cutting hours into minutes when winding either small or large motors. By the use of this machine, the repairman can have a complete set of starting or running coils ready in minutes. The tedious grind of feeding the wire through the slots is eliminated. The remaining operations consist of fitting the coils into the slots, inserting the wedges, and making the connections. Some points in favor of the prewound coils are: all coil groups are alike as to size and shape; errors in counting turns are practically eliminated; adequate clearance may be provided between coil ends; time is reduced; and the resistance of each coil is the same.

It is possible to purchase a winding machine from an industrial equipment supplier (call a local motor rewind shop to locate one) or one can be constructed in the home. The essential winding machine consists of a motor which is geared down to drive a shaft which holds a coil form at one end. The winder starts the machine and guides the correct amount of wire over the form as it slowly rotates. When the coils are finished, they are removed from the form and tied or taped. They are then ready to be inserted into the slots. A

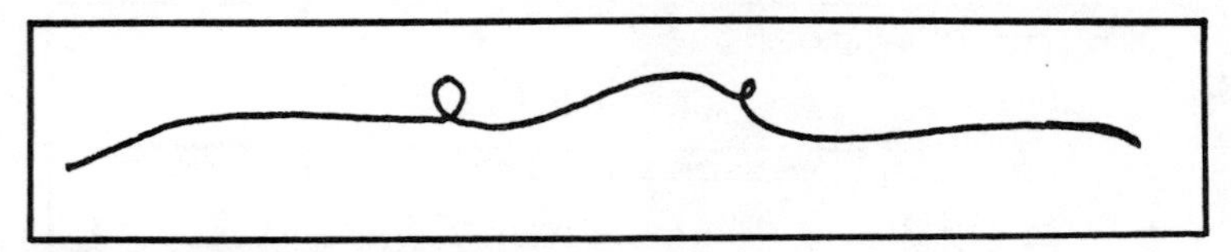

Fig. 3-6. Wire kinks are a no-no.

small, contact counter, which serves as a turn counter, enables the operator to wind any type of coil with accuracy.

A winding machine may be a large, free standing device, or it may be built into a bench, or it can be made portable, so it can be taken from one place to another. Figure 3-7 shows a bench or free standing coil winder in which all the hardware is kept under the table. Figure 3-8 shows another layout in which the entire mechanism is placed on a board which allows it to be portable. In both cases, a small motor, perhaps ⅛ horsepower is required. Assuming that it turns at 1725 rpm, it is necessary to reduce the speed to 30 or 60 rpm at the coil mold. Using a 1 inch pulley on the motor and a 12 inch pulley as the first reducer gives the shaft which holds this large pulley a speed of about 143 rpm. Using a 2 inch or 3 inch pulley on this shaft, connected to an 8 inch or 9 inch pulley at the mold shaft will reduce the speed to about 35 to 45 rpm.

In Fig. 3-7, a split shaft is shown, with a clutch located on it. This clutch is simply two discs which are pushed together to allow the machine to be stopped and started by foot, reducing the problem of the mold running out of control should the operator make a mistake while winding the coils. The only problem with this type of

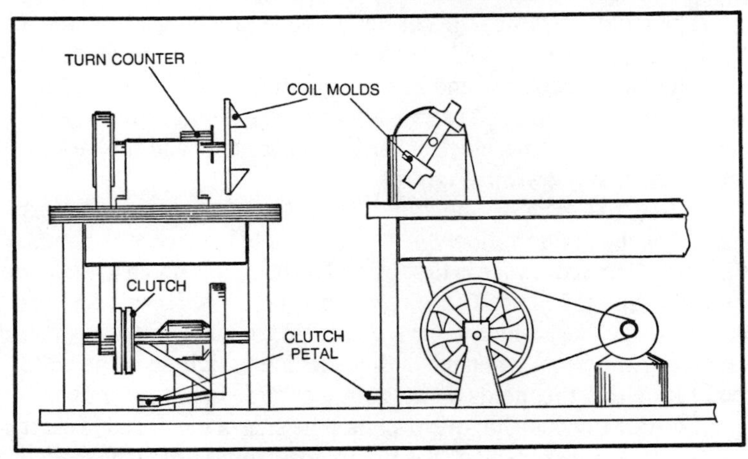


Fig. 3-7. A coil winding machine.

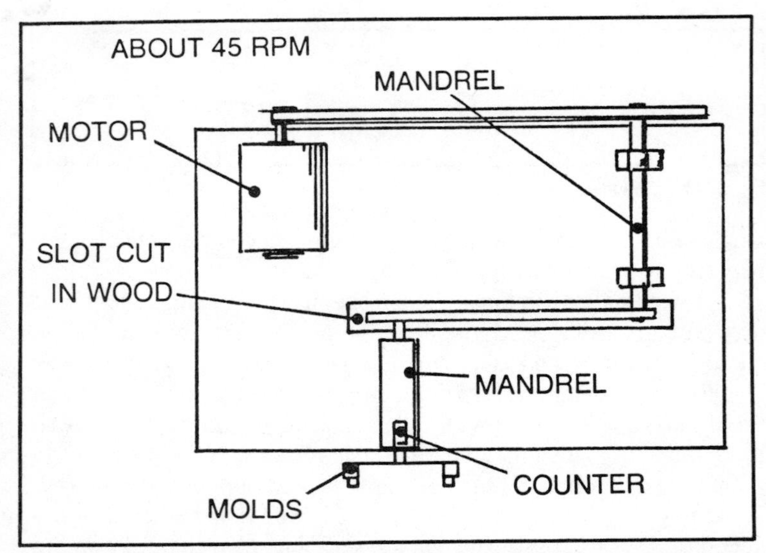


Fig. 3-8. Mounting the parts on one level. This platform is elevated with blocks for clearance of pulleys and molds rotating below its surface.

clutch is that it is impossible to obtain a smooth tension on the wire because the discs have a tendency to grab or slip. Clutch assemblies can be purchased, but are costly. The shafts themselves may be simple mandrels with sleeve, or pillow block bearings. The coil molds can be attached with a simple arbor. The turn counter counts the number of revolutions the mold makes and indicates the number of turns of wire wound on it.

The molds on which the coils are wound present no problem if the shop is equipped with a lathe. Molds to suit coils of any sizes can be turned out as required and a representative assortment of forms will soon be on hand to fit practically any type of coil. These molds should be turned from a block of hard wood, sanded, and shellacked. Such a mold is shown in Fig. 3-9.

A notch is made in the side wall of the largest and smallest channel on the mold to allow the wire to be held in place at the ends. When winding a coil, wire is first placed through this notch to keep it from falling off the mold. The wingnuts and bolts hold the two molds on the spacer bar, which allows the mold to conform to any coil sizes. It is important when using the machine to tighten these bolts securely so that the molds do not rotate or slide in the spacer bar.

To start a winding on the machine, the size of the old coils must be accurately measured so that the molds can be properly spaced apart. Too big, and the coils may hit the end covers of the motor, too small, and they will bunch up on the stator or not go in at all. Also, the width of each coil must be measured to allow the winder to select the right molds and the right channels on the molds for the coils.

After these measurements are made and the molds are securely fastened to the machine, the counter is set to 0. The wire is started in one of the notches, leaving a suitable length for connections. The motor is started and the winding begins, either first on the smallest channel and progressing up to the largest, or vise versa. It does not matter what direction the machine rotates, since the final polarity of the poles is determined by the way the coil groups are placed in the stator and how they are connected.

When one group of coils is finished, the machine is stopped. Taping or tying the wires between the mold blocks will keep the coils from unravelling when they are removed from the machine. The wingnuts are loosened and the blocks are pulled out, leaving the group in tact. Their appearance should resemble those shown in Fig. 3-10.

The remaining poles are wound in the same manner. When completed, they are placed in the stator slots and connected in the proper order of cw or ccw. It may become apparent right away that this machine requires cutting off the ends of each group as they are finished and setting them aside until all the poles are wound. In a new motor, the poles are wound with a continuous length of wire, with no breaks or cuts between the poles. Providing some means of

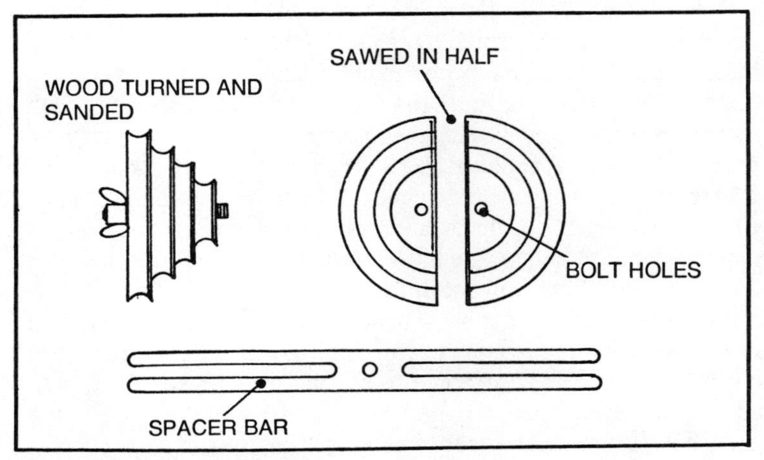


Fig. 3-9. One type of coil mold.

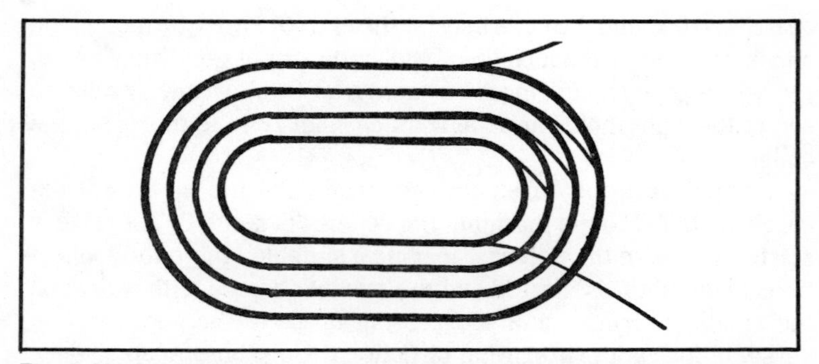

Fig. 3-10. A coil wound in a coil-winding machine. Note the even spacing between the coils.

attaching finished groups to a portion of the machine which would allow them to rotate along with the molds while the other poles are being wound would enable the use of a continuous wire, thus requiring no soldering or splicing. See Fig. 3-11.

## AUTOMATIC CLUTCH

An automatic clutch would be a great asset to the machine. Not a clutch which would automatically engage and disengage by itself, but one which would slip smoothly to keep a good tension on the wire as well as to be able to freewheel enough to allow the operator to grab and stop the molds with his hands should something suddenly go wrong (like one mold falling out of place). A fluid coupling is an ideal solution.

An automobile fan clutch is a device which will do just what is described above. The clutch transmits its power through a silicone fluid. Used fan clutches can be found in junkyards for a very low price, so they do not add much to the cost of the machine. The clutch must be mounted somehow on the intermediate shaft so it can run at the proper speed. A speed of 600-900 rpm at the clutch should allow it to transmit just the right amount of power for the winding. Since each clutch will be a little different, the operator may need to experiment. The pulley sizes, which will be different than those of the original machine, will determine the intermediate speed as well as the final speed at the mold. To mount the clutch, it can be bolted to a wooden disc in the same way it is bolted over the water pump in a car. The disc can be attached to a mandrel at its center. For an additional convenience, a pulley from the same car which has the same bolt pattern can be used to simplify the choice of parts. This is only one method of attaching the clutch. Figure 3-12 shows a

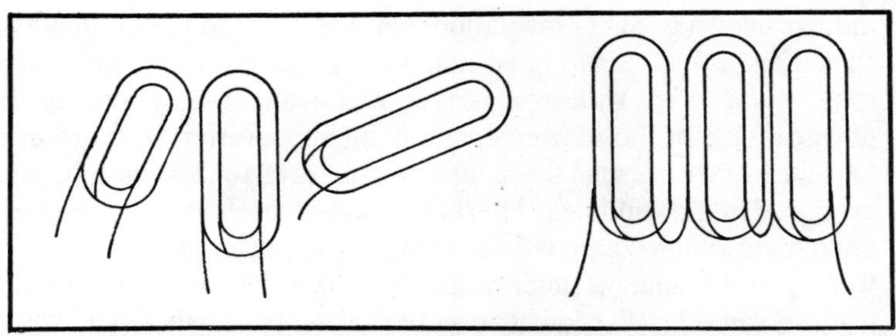

Fig. 3-11. On the left are finished groups which must be soldered together at their ends. On the right, they are wound from one piece of wire with no breaks.

modified winding machine with a fluid clutch and a block to carry finished poles to allow winding with one run of wire.

The type of wire used to wind the motor is very important today, much more critical than in the past. Several decades ago,

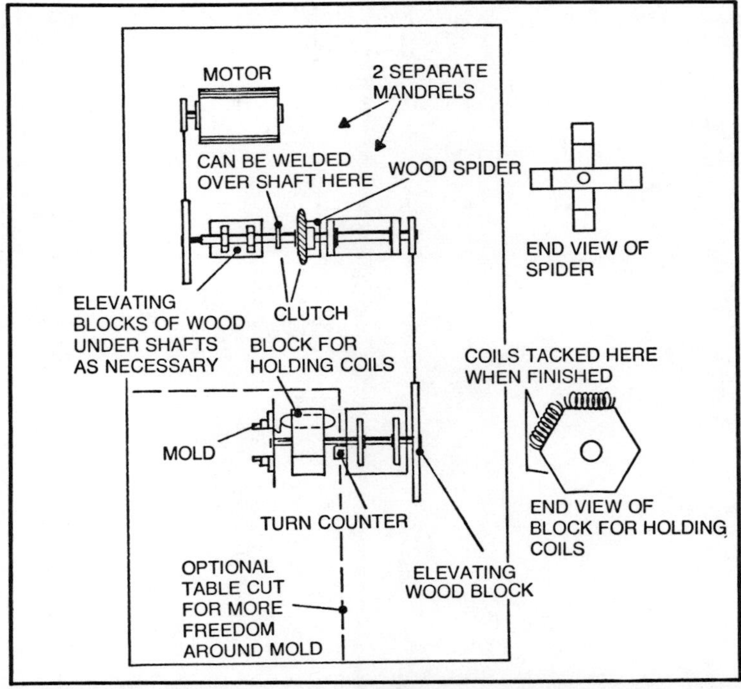


Fig. 3-12. Here, the input end of the fan clutch is welded to one of the mandrels. A second mandrel, directly in line with the first, has an X-shaped spider which engages the fins on the outside of the clutch so it will turn with the clutch. An elevating block of wood under the mold shaft raises the pieces up off the table will allow adequate clearance for working, as well as for pulley clearance.

motors used a grade of insulation that was allowed to get no hotter than 100-110° C. Since most magnet wire at the time was of that type, it was easy to use whatever was available. By making the physical size of the motors smaller, their operating temperatures became higher because there was less material to dissipate the heat produced in the windings. Therefore, higher grade insulations were developed which could withstand temperatures of up to about 180° C. When rewinding a modern motor, it is extremely important to use the same grade of insulation that was previously used, otherwise the motor may burn out even though the winding may be good mechanically.

Below are the current insulation classes and their maximum temperatures (all temperatures in centigrade). The insulation class is usually specified on the motor nameplate.

| | |
|---|---|
| A | 105° |
| B | 130° |
| F | 155° |
| H | 180° |

# Chapter 4

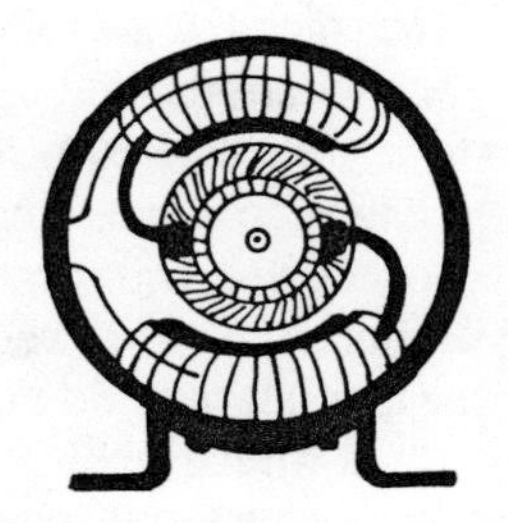

# Armature Testing

Much has been written about the winding of armatures but little about the methods of checking up on the work as it progresses. The winding of armatures is clean work and exceedingly interesting to those who like it until some form of trouble turns up. Nothing can ruin the day for an armature winder like a "bug" in the winding; one that stubbornly resists all efforts to locate it. Because he has been so close to the work, the winder himself may have difficulty in finding his own mistake; a mistake that may be obvious to a fellow workman. In the larger shops, the foreman, or another winder, may be called upon to do the trouble shooting, but the winder who works alone must depend on testing methods to insure accurate results.

In the usual course of winding, the winder goes about the task more or less automatically, having developed a certain familiarity with the work. Ever so often, however, an odd job will turn up; one that so taxes the ability of the winder that an error may result from his confusion. The error may be a short or a ground caused by trying to pack a large number of turns into a small slot. It may result from making a wrong connection or any number of other reasons. Most of the writer's troubles in this respect have come about as a direct result of telephone call interruptions or from the idle chatter of well meaning visitors.

Regardless of how careful the winder may be in his work, difficulties will occur which cannot always be prevented. The next best thing, then, is to detect such errors as soon as possible after they occur; not after the winding is all but completed. To do this the winder must use a system, a methodical routine of work and test, work and test. The worst troubles that may occur in armature winding are easily corrected *if discovered in time.* Each coil that is wound over a defective one makes the trouble shooting that much more complicated and the remedy more difficult.

In armature winding, the most frequent causes of trouble can be traced to the following defects: short circuits, grounds, reversed connections, and open circuits.

Short circuits are the most common defects encountered. They may occur between the turns of a coil, between two coil sides occupying the same slot, between coil ends where they lap over each other, between coil leads, and between commutator bars. Open circuits do not occur often. Grounds are also rather uncommon where good slot insulation and fibre end laminations are used, and where the winding is performed in a careful manner.

## TEST COMMUTATORS FIRST

The commutator can account for a lot of the winder's troubles if proper testing is not done in advance of the winding operation. As soon as the armature has been stripped, the commutator should be tested for grounds to the shaft and for shorts between bars. The test for grounds can be made by brightening the commutator and wrapping a bare copper wire around it three or four times so that contact will be made with all bars. Test from commutator to shaft using a voltage much higher than the normal operating voltage of the motor.

Small motor commutators of the 110-220 volt type should be tested at about 1000 volts for at least 60 seconds. A flash test is not always reliable because a little time is often required for the defect to show up. Alternating current is best for testing and can be stepped up to the voltage desired by means of a transformer. The commutators of very small motors, such as are used in sweepers, mixers, etc., are often unable to stand this test because of the thinness of the insulation, but such small commutators should be able to withstand double their normal operating voltages. A rule often followed in testing the commutators of larger motors, ½ hp or more, requires the application of 1000 volts plus twice the normal operating voltage. The commutator of a 1 hp, 220-volt motor would be tested at a voltage of 1440 volts, according to this general rule.

Figure 4-1 shows the method of testing commutators for grounds. A transformer can be constructed in the shop that will have taps for testing voltages of 100 to 1500, or more, in steps of 100 volts. The money and effort spent in constructing a good variable voltage transformer is well justified, and the outfit can be used for many other kinds of testing.

The commutator should be tested also for shorted bars. In ordinary work, this can be done satisfactorily with a 110-volt test lamp. The trouble most generally found consists of carbon deposits on the surface of the mica slot insulation. When such a condition exists, the test current arcs across these places and makes their detection a simple matter. This test is made easier by placing the armature in an armature stand so that it can be turned freely. Mark one bar as a starting point and apply the test points to each pair of bars as the armature is turned slowly. If much of this testing is required, the construction of a bar to bar testing device, such as the one shown in Fig. 4-2, will be justified. Such a device not only saves time but eliminates the chance of skipping a bar or two during the test.

Most shorts between commutator bars can be cleared up by cleaning the surface of the slot mica. An undercutting machine is best for this purpose, but the point of a knife, or a specially ground section of hack saw blade may be used. If the burned spot goes deep into the mica, it must be dug out, and the hole that results should be filled with some form of commutator cement. If no commercial form of commutator cement is available, a good substitute can be made by mixing plaster of paris and shellac into a thick paste. Pack this into the hole and test again after it hardens.

Several mechanical tests should be applied to commutators before the rewinding of the armature is begun. The commutators on armatures that have been burned in ovens should be given a rigid inspection for loose bars. Loose bars in a commutator may not show

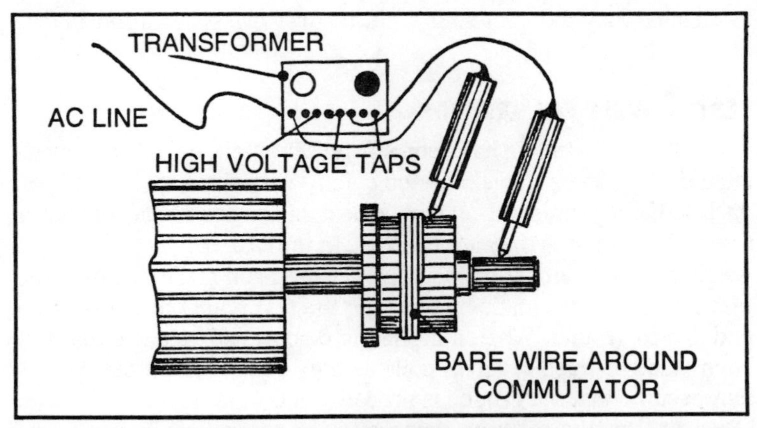


Fig. 4-1. A quick method of testing commutators for grounds is shown here. The wire connects all segments electrically so that only one test is required.

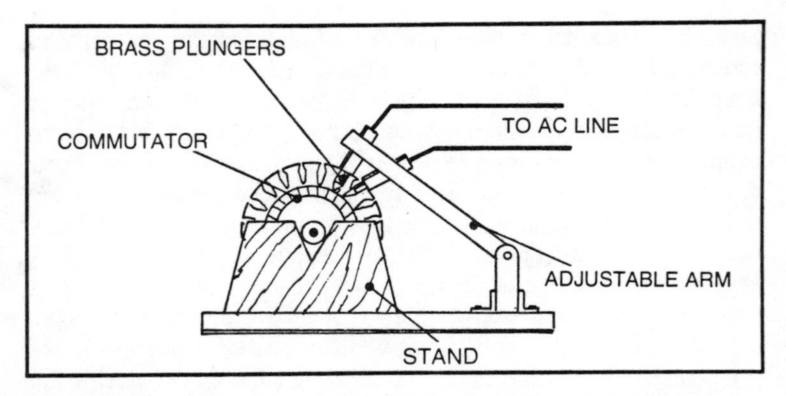


Fig. 4-2. This handy rack facilitates the testing of commutators for shorts between bars. Springs in the hollow plungers, insure good electrical contact with bars.

up even when the commutator is stripped or dressed in the lathe, but they may become evident when "staking in" the coil leads. Gentle tapping with a light hammer will cause loose bars to shift their position. Figure 4-3 shows how bars may shift on both the horizontal and vertical type of commutators.

If there is any indication of looseness, the commutator should be repaired or replaced. Nothing can cause more trouble than a poor commutator. Some fractional horsepower, single-phase motors now have commutators that are insulated between the bars and a bakelite compound instead of mica. These bakelite strips frequently burn and char when overheated and can seldom be used again. A winding is no better than its commutator and the wise winder ascertains that the commutator is in first class condition before starting the winding.

## TESTING COILS FOR GROUNDS

After the winding has been started, frequent tests for grounds should be made on the completed coils. Tests need not be made on each coil as it is finished but should be made on each group of four or five coils as soon as they are in place. In the case of the straight loop winding, where a continuous wire is used from start to finish, the test for a ground is made by touching the test points to the starting end and to ground. When a ground is discovered, and the test has been made on each group of coils as they were wound, the winder may assume that the ground is probably in the last few coils wound. Occasionally, the tamping down of an upper coil will result in a ground in a lower coil. The exact coil in which the fault lies may be located by cutting the loops between coils, separating the winding into sections. The section of the winding containing the ground can then be separated into individual coils and the test points may be used to locate the faulty coil. Figure 4-4 shows a method of locating a ground in a loop winding.

When the armature is wound with two or more wires in hand there will be free ends projecting from the slots, and each coil may be tested at any time. It is a wise plan to wind two or three coils and then test them before proceeding with the work. The upper, or finishing leads, can be bent back over the core to get them out of the way, and the test can be made from ground to each of the projecting starting leads.

Many grounds show up only after the wedging operation. The pressure of inserting the wedge in a tightly packed slot may result in a puncture of the slot lining, especially, if no fibre end laminations or heads have been used. For this reason, the winding should be tested again for grounds after the wedging operation and before any connections are made to the commutator. This may sound like a lot of extra work but it takes only a very few minutes and may save hours of time and the vexation of having to do a large part of the work over.

Nearly every winder has a pet system of his own for connecting the coil leads to the commutator. Some winders use one method on one type of armature, and another method on another type. The leads must be connected to certain commutator bars, of course, but there are different ways of handling them. Some winders connect the starting leads of each coil as it is wound to the commutator. The principle advantage of this method is that the winder has no trouble whatever in distinguishing his finishing leads as they are the only

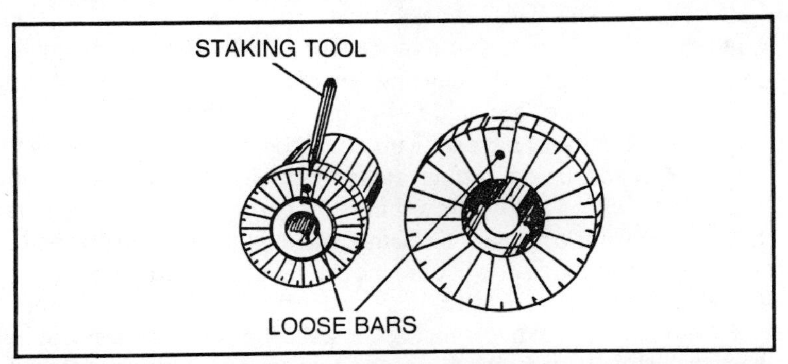


Fig. 4-3. Loose bars may be found in commutators that have been overheated. Test commutators mechanically before beginning the winding and avoid difficulty later.

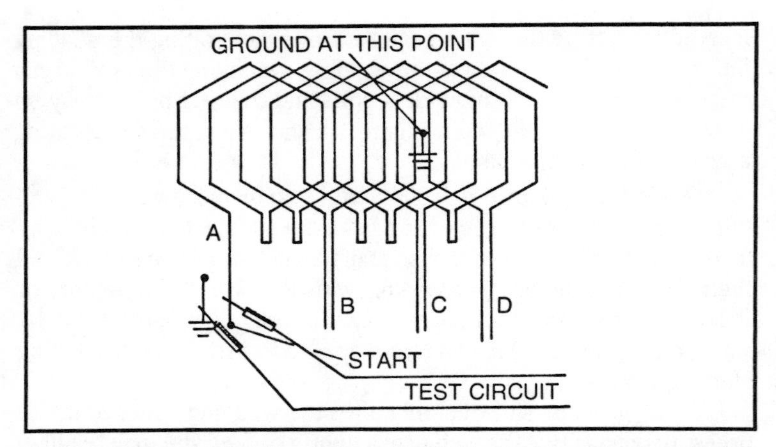


Fig. 4-4. A ground in a loop winding may be located by separating coils into groups as shown above. By a process of elimination, the faulty coil may be located.

ones that remain unconnected when the winding is completed. A disadvantage of this method is the covering up of starting windings. If an error has been made, it will be difficult to correct without unwinding.

Some winders use a natural winding method that leaves the starting leads at the bottom of the coil and the finishing leads on top. There is little opportunity for error in this system but there is a greater chance of trouble developing later, such as shorts and grounds. It is much more difficult to insulate leads properly that are buried deep under the winding. Other winders manage to bring out both the starting and the finishing leads of the coils in the top of the slots, even though the coils to which they belong are wound in the bottom of the slots. This is done by leaving these ends outside of the slot to which they belong until the rest of the winding for that slot is in place. These leads, the starting end of one coil and the finishing end of another, are then laid in the top of the slot.

The winder may have difficulty in recognizing start leads from finishing leads unless some sure method of identification is employed. One of the best systems of identification requires that the start and finishing leads be of unequal length. If the starting ends are clipped just long enough to reach the right commutator bars and if the finishing leads are left at least two inches longer, mistakes are not likely to occur. When winding consists of more than one wire in hand, which is usually the case with this type of winding, colored tracer wire will simplify connecting and will save the time of testing each circuit.

## MAKE COIL TESTS BEFORE SOLDERING

Tests for shorts, grounds, and reversed connections should be made as soon as the leads are staken in the commutator slot, and before the leads are soldered. Unsoldered leads can be changed easily if a wrong connection is discovered. They can be raised if special tests should be required. All armatures can not be tested in the same way or with the same equipment. They can be given a final test for grounds, however, by applying the proper test voltage to commutator and shaft.

Armatures can be tested for short circuits in a number of ways. Dc armatures and some ac armatures can be tested quickly and accurately in an armature growler. Some of the ac armatures of the repulsion-start, induction-run type have cross connections between opposite commutator bars, and can not be tested properly in a conventional type growler. There is a special form of external growler available, however, that can be used with these armatures.

Short circuited coils in armatures of the type that can be tested in the regular growler may be found by passing a hacksaw blade above the armature, while the armature is revolved slowly in the growler field. A sticking or vibrating blade indicates the presence of a short in the slot beneath. Armatures that pass this test may still have trouble in the form of an open circuit or in reversed connections.

One of the best methods of testing for open circuits in a growler is by means of a small lamp connected to two test prods. A

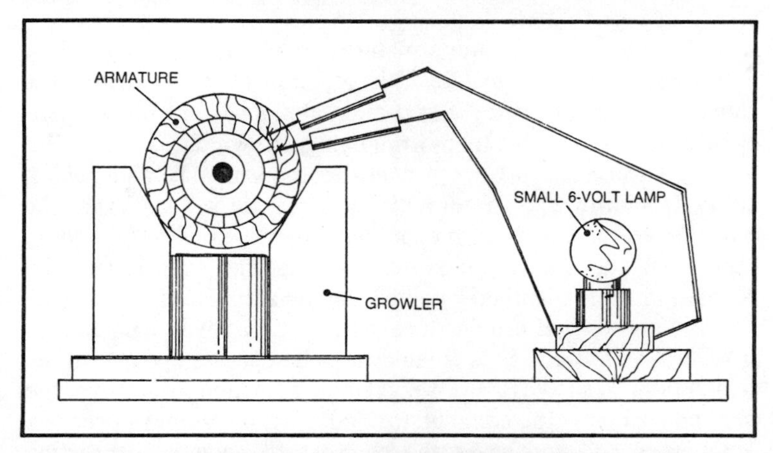


Fig. 4-5. A test lamp such as the one shown here provides an excellent means of locating shorts, grounds, and open circuits. Induced voltage from the growler lights the lamp.

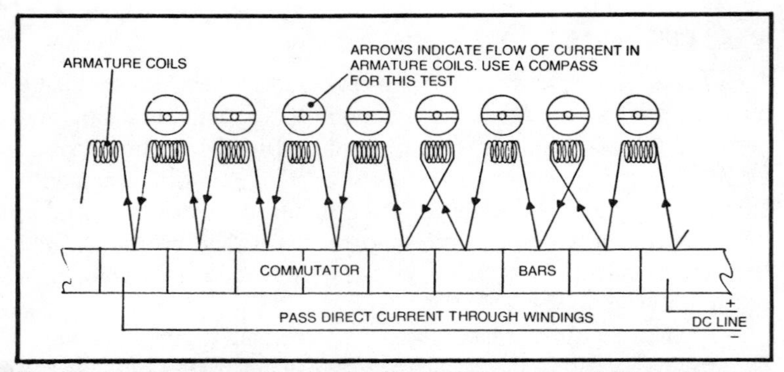


**Fig. 4-6. A small compass may be used to check for reversed coil polarity. In the case of smaller motors, it may be necessary to lift commutator leads during the test.**

good lamp for this purpose is a No. 63, such as is commonly used in automobile dash and tail lights. As the armature is slowly revolved in the growler, the two leads to the lamp are placed on adjacent commutator bars at a position where the lamp will light with the greatest brilliance. After this position is once found, all bars are tested in the same position. The test points may be held still while the armature is turned. Since the potential between adjoining bars is usually low, there is no danger of burning out the lamp. As a matter of fact, the voltage that will be induced in most armature coils will light the lamp rather dimly. Poor contact between the leads and commutator bars will cause the light to burn dimly while an open circuit will give no light at all. Equipment for making such tests is shown in Fig. 4-5.

Dc motor and generator armatures are often tested inside their field frame if it is available. This test can be made before the commutator is soldered if the commutator leads are bound sufficiently tight to prevent them from being thrown out of place. For testing purposes it is best to energize the armature and field of either a motor or a generator with a voltage that will just cause the armature to turn. A sensitive ammeter in series with the energy supply will indicate reversed coils, shorts, and open circuits by fluctuations of the ammeter pointer and by arcing at the brushes.

A compass will provide a reliable test for reverse connected armature coils. Current is applied to well separated points on the commutator as shown in Fig. 4-6. A compass is then passed over the coil ends opposite the commutator end and the readings observed. A reversed coil will cause the compass to point in a direction reversed from that observed for the properly connected coils. On small armatures, it may be neccessary to test one coil at a time to get an accurate reading. This is accomplished by lifting the coil leads from the commutator during the test, making certain each pair of leads are connected properly.

Cross connected ac armatures, and others of the repulsion-induction type of motor, can be tested satisfactorily without elaborate equipment by installing the armature in the stator with the brushes removed. With the bearings properly adjusted to allow no rubbing, connect the motor to an ac circuit and turn the armature by hand. A clear winding will allow the rotor to be turned easily while a short circuit at any point in the armature winding will cause the rotor to lock in one position. As stated previously, the brushes must not touch the commutator during this test.

Many short circuits, and a few grounds, that are found in rewound armatures, are caused by the soldering operation. For this reason, it is advisable to make the main test after the commutator connections have been completed and before they have been soldered. If trouble becomes apparent after the soldering job has been accomplished the winder will know that the source of the troubles is in the commutator, and not in the winding proper.

Shorts that appear in the commutator after soldering can usually be traced to three things: solder between commutator bars, solder that has flowed below the winding and shorted commutator leads, and the use of acid flux. Small particles of solder often wedge in commutator slots and complete an electrical path between bars. These will sometimes burn out in the growler test, but if not, they can be located by using the method shown in Fig. 4-5. Test from bar to bar and the trouble will be located between bars that do not cause the lamp to light. Surplus solder that has found its way below the winding and caused a short can be located in the same manner. If an acid flux is used in the soldering, the surplus acid will sometimes soak into the insulation and cause shorts or leaks. The possibility of this kind of trouble should be avoided by using a rosin flux.

If the job is worth doing, it is worth doing right. Frequent tests while the work is in progress will result in better work and will eliminate the need for patch work on mistakes made during the earlier stages of the winding.

## Chapter 5

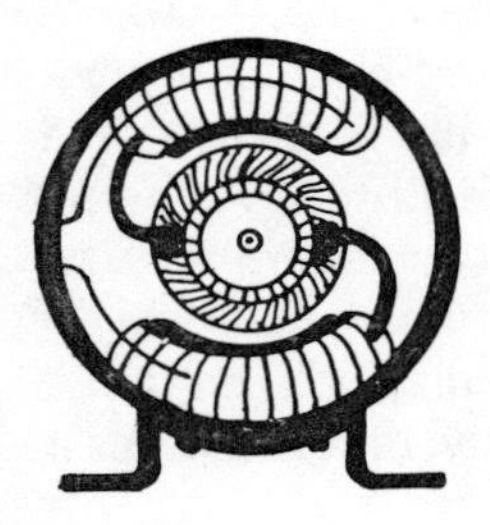

# Equipment for Testing Stators

After a stator is rewound, it is always tested in several ways to make certain that the new core is good. Motors brought in for tests are also checked in the same ways to pinpoint defects and allow the winder to correct local faults without rewinding the whole primary. There are two classes of motors coming into shops for repair, those of the small, single phase type, and those of the larger polyphase type.

The majority of the motors are single phase, being used for domestic appliances and tools. Experienced winders find little trouble in diagnosing these motors, spending little time or thought about what they are doing. Single phase windings are simple and easy to work on, often requiring only a polarity check and one or more tests for shorts and grounds. To those not so experienced, the following methods can be used.

The first area of testing covers motors which are brought in for repair. A quick visual inspection will reveal physical damage. Further probing will reveal such things as locked rotors, damaged bearings, etc. These mechanical problems can be solved by servicing the proper parts. The next step is to open the motor and look at the windings. Coils which are melted and burned are in definite need of replacement, but these appearances are sometimes subtle and not always certain. The motor can be given a test run. If it blows fuses or circuit breakers, one can assume that the winding is shorted. If, while under test, the motor does not turn at all, the fault may lie in the starting circuit. If it does start, or begins to rotate, the winding again is probably shorted if it trips a breaker.

If a starting winding is suspected of being open, it can be tested with an ohmmeter. Taking the motor apart and connecting a meter across the ends of the starting winding should give a reasonably low resistance. If the reading stays at infinity, there is a break somewhere in the winding. The running winding can also be tested in the same way for an open circuit.

Assuming that there is an open circuit in one of the windings, the next step is to find it, without ripping all the coils apart. If the motor has four poles, as shown in Fig. 5-1, an ohmmeter can be connected at one lead and the other probe touched to each pole at a spot where a tiny bit of insulation has been scraped off the wires. Starting at the pole to which the other lead is connected, progress around the stator until the meter reads an open circuit. At that point it will be noticed which pole has the break in it. The open pole can be replaced, being wound with the same gauge wire in the same fashion as the defective one. When completed, the winding should then have continuity between the leads providing that there are no more breaks elsewhere.

## CHECKING FOR GROUNDS

Grounds, or connections between the windings and some part of the frame can be checked with a multimeter. Setting the device for the highest resistance, one lead is attached to the end of the winding, while the other is touched to the housing, as shown in Fig. 5-2. A reading of several hundred kilohms or less indicates a ground, and in most cases, more than this also represents a potential hazard. If the winding is generally burned evenly all around, there may be several ground paths, in which case the whole winding should be replaced. If, on the other hand, the winding appears good, applying a high voltage from a transformer (several times the operating voltage at low current) to the leads will cause the point at which the ground is formed to smoke or flash. If it is only one pole, that pole can be replaced.

When a winding is visibly burned, it is almost certainly shorted as well. A short occurs when the insulation on the wire melts and allows the conductors to touch. Shorts can be the cause of fuse blowing in the run test, as well as smoking or burning while the motor is operating. Naturally, if the whole winding is bad, it must be replaced, but if only one part appears defective, it can be replaced separately. One way to test each pole for shorts is to take the motor apart and apply a reduced voltage to the primary. Using a hacksaw

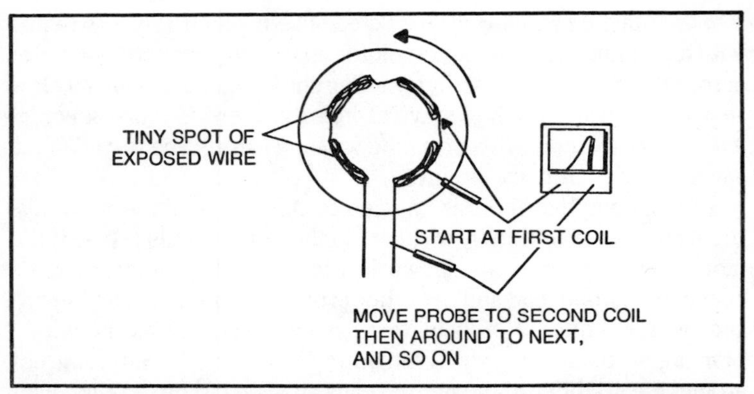


**Fig. 5-1. By scraping a whisker of insulation off each pole group and testing each one with an ohmmeter, it is possible to locate open coils. Be careful when scraping insulation not to damage the wires.**

blade, knife, screwdriver, or any other such metal object, bring the tip of it to the center of each pole on the inside of the core. The pole, acting as a magnet, will attract the tool. The pole which pulls the tip with the least amount of force is shorted. If there is more than one shorted pole, there will be varying pulling forces obtained at each group and hence the whole winding should be replaced.

Once the necessary repairs are made on the motor, it should then be tested before the new winding is dipped, so new problems can be fixed without the need to "chisel" out a freshly varnished winding. The same ground and short tests should be made to verify that a new winding is indeed good. The motor should then be given a test run, first at a reduced voltage, and then at full voltage. If it operates properly at the full voltage, it can be dipped and baked to complete the job.

There are other parts of single phase motors which require attention besides the stator. These include centrifugal switches and/or capacitors. The contacts which are operated by the switch must be free and undamaged. Sometimes, when a motor fails to start, the problem is a broken contact on the inside of the end cover. Capacitors can also go bad, either by opening up (a break in the metal conductors inside the unit), or shorting, in which case the dielectric material breaks down allowing the two conductors to touch.

## CHECKING CAPACITORS

Capacitors can be tested with an ohmmeter. If a meter, set on its highest scale, is connected to an electrolytic capacitor, the needle will rush toward the 0 mark initially and then gradually fall back towards the infinity mark as the capacitor charges. If it does not give any reading, the capacitor is open. If, on the other hand, the needle stays toward the 0 mark without falling back, the capacitor is shorted. When replacing a capacitor use the same size as the original, or if that is not available, one that is a bit larger.

Sometimes, motors other than the basic single phase types are brought in for repair. These may be complicated dual-voltage, multi-speed motors, or larger polyphase motors, each with many more poles and coils than the basic motors described earlier. It is very beneficial to have proper equipment for testing these motors as well as for their rebuilding. Sometimes, a motor with a burned winding is in need of repair. The windings can be traced from the original and duplicated for the new winding. Rarely, a bare frame is in need of winding, and consequently the winder must design a winding from scratch. Knowledge of basic design is essential when creating such a winding, and knowledge of testing procedures is essential to verify that the new winding will indeed work.

Rewinding a large polyphase motor is something completely different from rewinding a single phase motor. Single phase windings are so simple that there is little chance for error, especially for the experienced winder. With large polyphase motors, however, there is a great number of chances for error. This does not mean that one should be discouraged from polyphase rewinding, the task is rewarding and beneficial to the service man's knowledge. There are two keys for doing these big jobs successfully: planning the entire winding and operation out in advance, and thorough testing afterwards to catch mistakes.

Before ripping out any old winding, a detailed sketch of the coils and connections should be made. With newer motors which

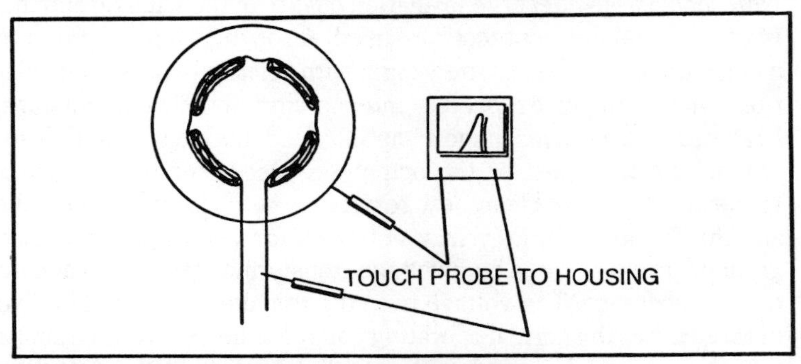


Fig. 5-2. Checking for grounds with an ohmmeter.

often are wound with continuous lengths of wire, or with connections that are hidden, making this diagram may require peeking and poking around the old coils with a small screwdriver. The connection diagram on the motor nameplate will be of some help. In this way, even if the particular winding is new to the winder, he can still recreate it by use of the sketch. The preparation of the diagram may be considered as one of the most important parts of the job.

After rewinding, the next step is to check the connections and check the diagram again. Often, a second person is best for this since it is easier for him to spot errors which the original winder does not realize he is making. In some of the larger shops this checking is done by the foreman or head winder. At home, or when there are no other servicemen around, a fresh look at the diagram is the best way to correct any flaws.

Each winder will have his own system of diagramming out windings, but these following points should be included in each: number of turns per coil, number of coils, number of poles, beginning and end of each phase, polarity of poles, direction of coil spooling (ccw or cw), and any other pertinent information. The connections between each coil should be clearly indicated.

If the winding passes inspection, it is then subjected to tests. Like the single phase winding, it is tested for grounds and shorts, but polyphase motors should be given extra attention to polarity and phase balance. Polarity refers again to the cw or ccw spooling of the coils and the phase balance test is used to verify that there is an equal amount of current flowing in each phase while running, and that one is not being overloaded or weakened.

Figure 5-3 shows the test for grounds with polyphase motors. These are nearly identical to the techniques used in Fig. 5-2. An ohmmeter can be used in place of the test lamp if one wishes. Different factors affect the insulation quality of the winding and this is why several different tests are used. An open, dripproof motor in a clean environment may be wound with relatively passive insulation, while a motor exposed to intense vibration, acids, moisture, heat, etc., requires much more durable (and more expensive) insulation. The lamp test or the ohmmeter test is for fairly obvious grounds. These are clean, low resistance paths between the coils and the frame. Still, they may not work for very high resistance grounds. In this case, a high voltage transformer test or a magneto test is necessary. The voltage from the transformer should be two or three times the normal operating voltage. The high voltage jumps through any small grounds with resulting arcing at the defective

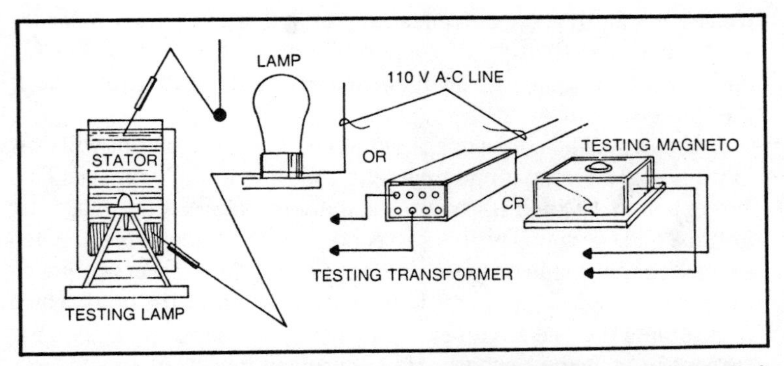


**Fig. 5-3. Rewound stator windings should be given a thorough test for grounds. A lamp test followed by a magneto or high voltage transformer test gives assurance of clear coils.**

spot. This high voltage test should be brief, or else the high voltage, if left on too long, may burn through good parts of the winding and make, rather that indicate grounds. A crank type magneto can be used in place of the transformer, but this method is losing popularity because the output voltage varies with the crank speed, hence there is no definite way of making a certain test.

It is important to test all electrical paths inside the motor. In dual voltage motors, there will be several, where as in a single-voltage, single-phase motor there may just be one. Each phase inside of a three phase motor is connected to the others in most motors, so only a few tests need to be made, but it is best to check for grounds from *all* the leads to be certain. Figure 5-4 shows some of the different load connections for three phase motors. Note how some coils are independent from the others.

The screwdriver pulling test described previously is one way of detecting shorts, but another method more commonly used with polyphase motors is the use of an internal growler. The growler is a small coil of wire very much like the core of a transformer, which is moved around on the inside of the stator. If the growler passes over a shorted coil, the current flowing in the growler increases. This increase can be detected with an ammeter.

Many service men have only a hazy idea of the working theory of the growler, and hence a brief description may be of some benefit. Almost everyone who has worked with electricity to any extent understands the operation of a transformer. Also, most repairmen understand the working principles of the induction motor, and how current consumption is influenced by load. We know that when alternating current is passed through a coil of wire wound around an iron core that this core becomes magnetized. If a second coil of wire is brought near the first, a voltage is induced in the coil which is directly proportional to the voltage in the first coil and to the ratio of the number of turns in the two coils.

If that second coil is not a complete circuit, no current will flow in it and only a very minute current will flow in the primary. This current is that which is required simply to stabilize the magnetic field around the coil. This is exactly how a transformer works. When a connection is made on the secondary so that the current is allowed to flow through it, it in turn generates a magnetic field which counteracts the one set up by the primary. To make up for this, the current flow in the primary must therefore increase.

## USING A GROWLER

If it is possible, obtain a small transformer and connect an ammeter in series with the primary. While the transformer is idling, the ammeter will show only a trickle of current flowing. Connecting a load to the secondary will give an immediate increase in current to the primary. The growler works the same way, although the unit itself is essentially only the primary, and the stator windings of the motor are the secondary. If the growler is held over a good coil,

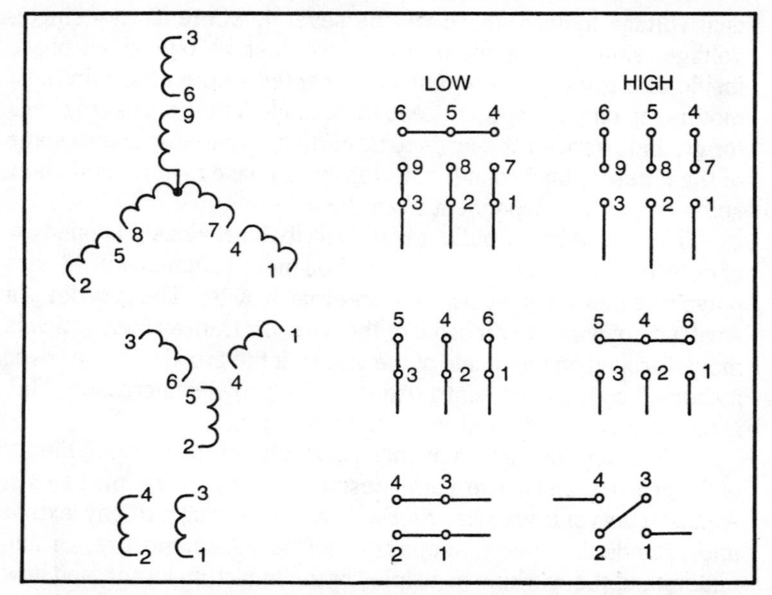


Fig. 5-4. Dual voltage motors have independent coils inside, each of which must be tested separately for grounds.

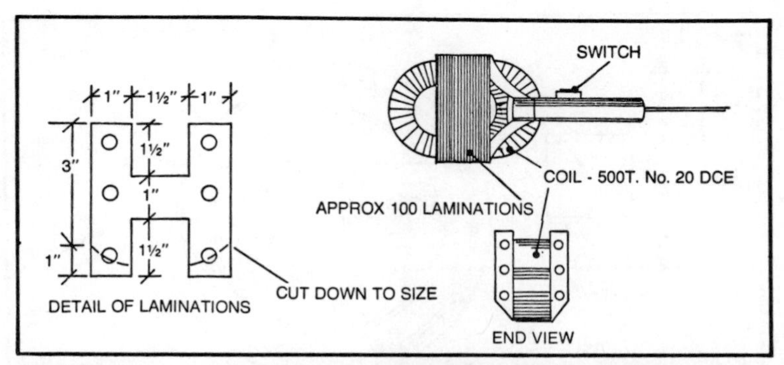


Fig. 5-5. Short circuits in stator windings may be located easily with an internal growler. This diagram gives construction details for a 110-volt, 60-cycle, ac growler.

there is no circuit in the secondary, so there will be little current flowing. If the coil is shorted, however, the current will increase. Figures 5-5 and 5-6 shows the construction of a homemade growler and its use.

To make the growler, an old transformer is an ideal place to start. Remove the old windings and using a hacksaw, cut the laminations so the remaining block is H shaped, as shown in Fig. 5-5. These laminations will tend to rip apart when sawed, so it is best to do the sawing with the block clamped in a vise. The bottom legs should be rounded so as to fit the inside bore of the stators. The wire can be wound onto the core and covered with tape. The laminations can be bolted together and attached to a handle for easier maneuvering. A thumb switch on the handle will make testing easier.

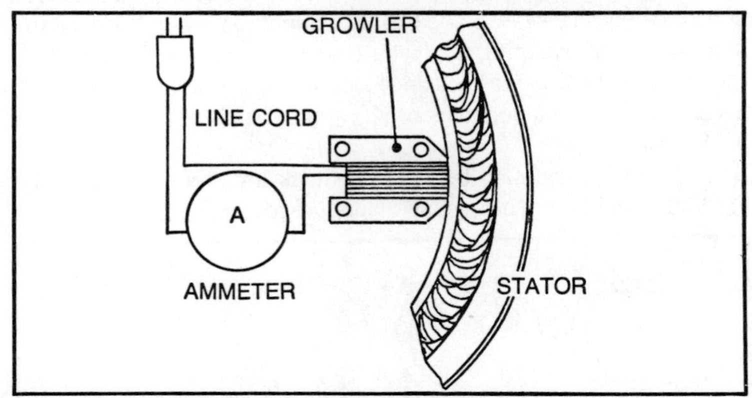


Fig. 5-6. Shorted coils form complete circuits, which draw current through the growler.

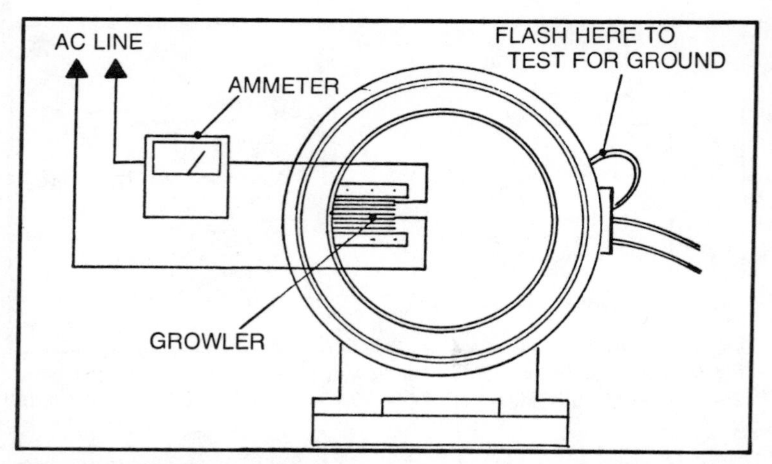


Fig. 5-7. The growler, in conjunction with an ammeter, can be used to locate either grounds or short circuits in stator windings. Grounds may also be located by the flash test.

The growler can now be used to find shorts in the following manners: in Fig. 5-6, the growler is connected to an ammeter and plugged in. Starting with any spot on the inside of the stator, the growler is moved slowly around the core. The current reading on the meter will be fairly low until the growler passes over a shorted coil, at which point the current will increase. If the coil looks burned, it should be completely replaced, but if it is a new coil just put in place, chances are that the short is only in one spot. The next is to find out in which slot the short lies.

The growler should be placed over one coil side of the shorted coil and then a hacksaw blade can be brought to rest on the opposite side of the coil. Keeping the distance from the growler to the blade the same, move the two back and forth over a few slots until you find a spot where the blade sticks or vibrates. The short will be located in that spot.

A growler can also be used to check for grounds. While the growler is being moved around inside the stator, each of the motor

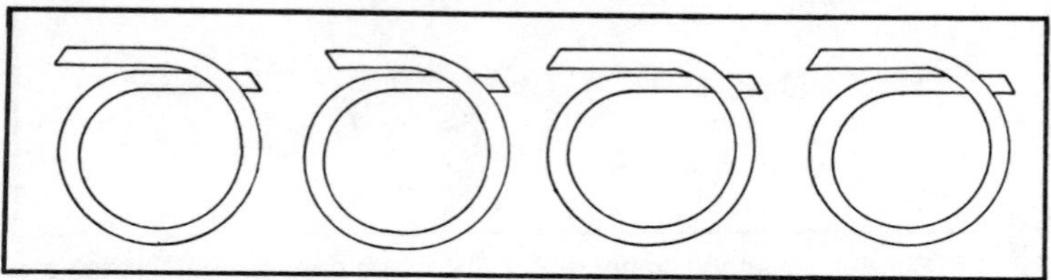

Fig. 5-8. Four identical coils.

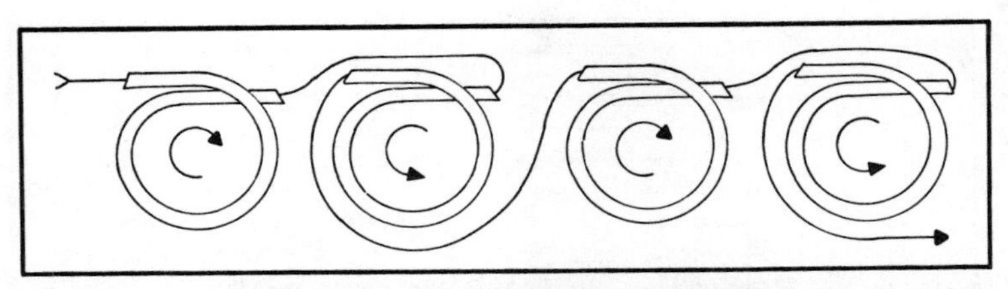

Fig. 5-9. Although these coils appear to be identical, the manner in which they are connected determines their actual polarity.

leads can be touched to the frame. If there is a ground in any coil, touching the lead to the frame will form a complete circuit and current flow in the growler will increase. See Fig. 5-7.

Occasionally a mistake is made in connecting the coils in polyphase motors. Consider the rolled out, single-phase winding in Fig. 5-8. Which way are the coils "wound", cw or ccw? There is no real answer to this question, because the way that they are wound depends on how they are connected from one end to the other. To make a four pole winding, with each coil wound in an opposite direction from the next, they could be connected as shown in Fig. 5-9.

A simple polarity test to determine proper connections is the compass test. In this method, a low voltage dc source is connected to each phase of the winding and a small pocket compass is moved around the inside. As the compass passes over each pole, it aligns itself with a field of the coils. By noting which direction the compass points it is possible to mark the polarity of each coil to verify that the connections are correct.

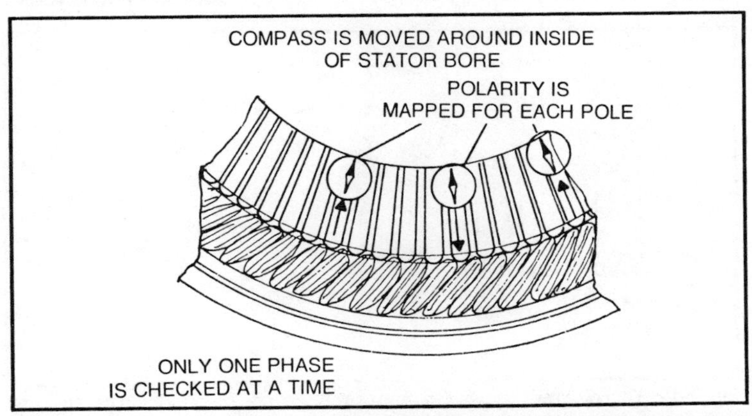


Fig. 5-10. A pocket compass is useful in checking the polarity of a phase winding. The test must be made with a dc voltage applied to the winding.

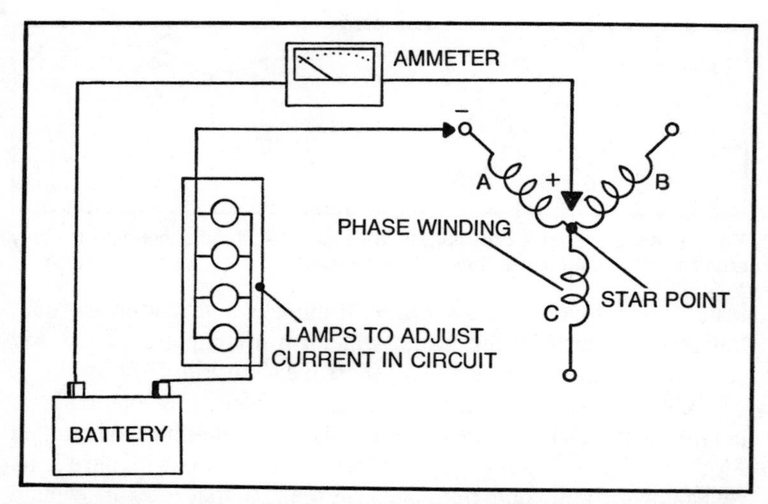


Fig. 5-11. In testing a three-phase star winding for polarity, one of the dc test lines must be connected at the junction of all three phases and the other to the phase under test.

A six volt heavy duty battery can supply enough current for this test to work on all but very large motors. The current flow should be about 5% the full load current through one phase. To effectively control the current flow, the battery is fed through a network of six volt lamps wired in parallel. By inserting different numbers of lamps, the current can be made large or small as desired. An ammeter in the circuit gives a reading of how much current is actually flowing. An alternative to the battery and lamp arrangement is an expensive variable power supply, but either will work.

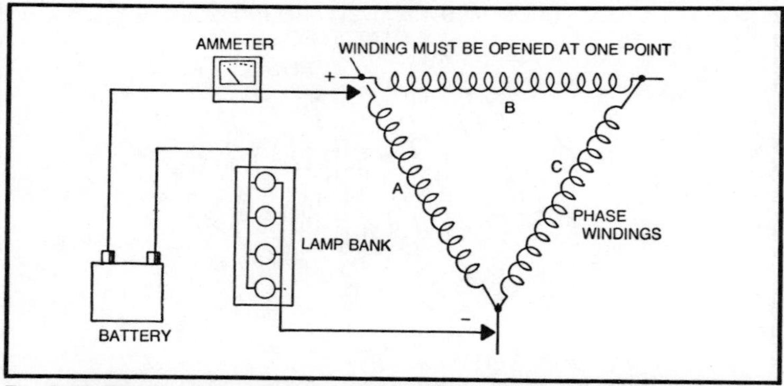


Fig. 5-12. This diagram shows the proper connections for making a polarity test on a three-phase delta connected stator. The winding must be disconnected at one point as shown.

When testing for polarity, only one phase is tested at a time. The leads are connected to the motor and the compass is moved around the inside. If the north end of the compass points toward a pole, the pole is called a north pole and is marked with a crayon in an arrow pointing one way. If the pole is a south, the arrow points the opposite way. The next step is to test the remaining phases in the same way. Figure 5-10 shows how the compass points with different poles.

In testing a three phase star connected winding, one of the dc test lines must be attached at the star point. The other test lead is alternately connected to the phase leads A, B, and C. Use the positive lead for the center connection. This is shown in Fig. 5-11. If the winding is that of the dual voltage kind, connect the leads for the low voltage and perform the test, then repeat the test at the high voltage. Remember to keep the same battery lead to the center at all times (positive, in the case mentioned above).

If the motor has a delta primary, one connection must be opened up at the junction of two phases, as shown in Fig. 5-12. This is so that each phase is tested as an open circuit. Unlike the star winding, in which one lead remains stationary, both leads must be moved in a delta primary to reconnect to each phase. The proper order in which the leads are moved is shown in Fig. 5-13. If this order is not done correctly, the results will indicate that even a good winding is bad, because the reversing of the current in one of the phases will throw it out of place from the others.

The finished result for either type of winding would look something like that shown in Fig. 5-14, showing polarity in each phase of a four pole primary. Any deviations from this order would

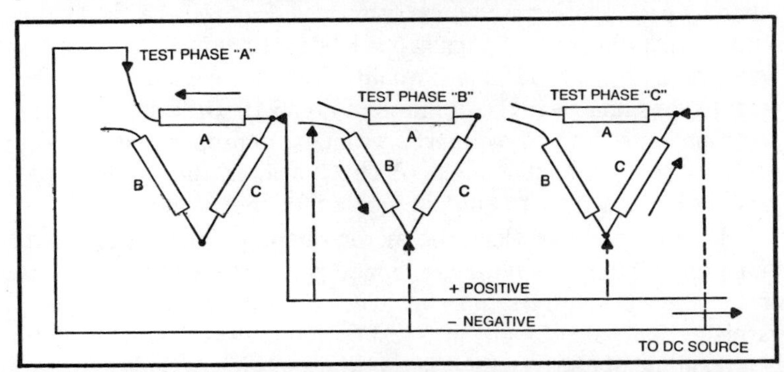


Fig. 5-13. In testing a delta connected winding, each phase must be checked with the current flowing in the same direction. The proper battery connections are shown above.

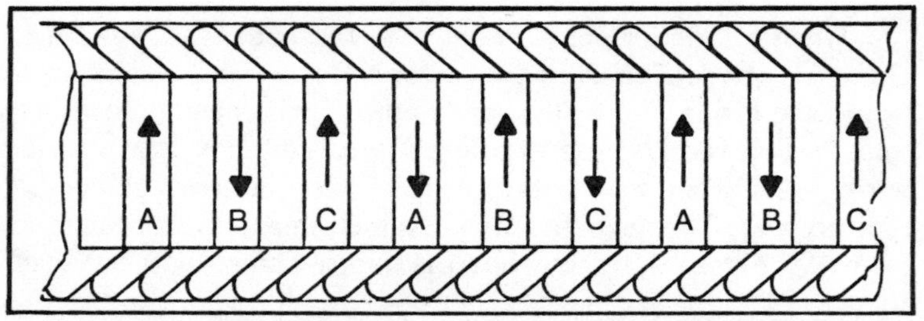

Fig. 5-14. A three-phase winding is projected on a plane surface here to show the results of a polarity check obtained with a compass. The phases are indicated as "a", "b", and "c."

indicate an improper connection between two coils or an improper connection of the motor leads.

Each phase should have the same resistance and impedance as the others, so that the current flowing in each will be the same. One way to check for phase balance is with a highly sensitive ohmmeter. If one has on hand such a meter that will measure in tenths of an ohm (such as a digital meter), it can be connected to the primary in the exact same ways that the six volt test supply was used previously to check polarity. This goes for both star and delta primaries. Ideally, each phase should be equal. Things that account for phases not being equal in resistance are: differing numbers of turns of coils in each phase, or shorts, which should have been previously detected.

A better way to check for phase balance is to use a variable transformer to supply a reduced voltage to each phase. By installing an ammeter and a voltmeter to the circuit, it is possible to monitor the power taken by each phase. The transformer should be set to give about one-quarter the normal voltage to the motor. The current taken by each phase can then be compared. Figure 5-15 shows the setup with a star primary. Again, one lead is left at the center. If this was a delta primary, one phase would have to be opened up, and the tests made similarly to those shown in Fig. 5-10. Differing readings could be caused again by shorts, grounds, improper coil connections, or reversed connections. A check against the winding diagram is the best way to start in looking for the trouble.

If all tests prove okay, the motor should be given a test run. Since the winding has not been dipped yet, it is normal to experience some vibration, but this should cease once the motor has been varnished and baked. A physical test for loose coils within the slots is a good idea before certifying that the motor is sound.

It is not always possible to test run motors in the shop because of the lack of the proper power supply lines. Large motors with high

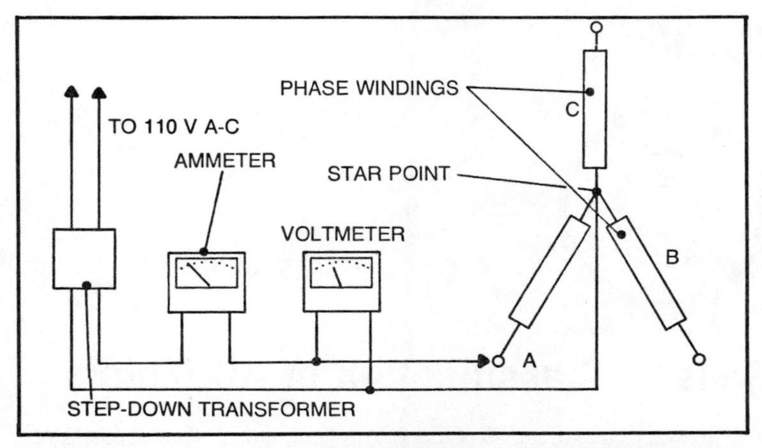


**Fig. 5-15. An important test on every three-phase stator is the phase balance test. In this test, a known voltage is applied to each phase winding and the current measured.**

starting currents cannot be started on small capacity lines, but the use of the variable inductors described in Chapter 1 allow large motors to at least be powered up at idle without straining the line. This is not the same as full load operation, but in most cases, if the motor tests good, it is not necessary to actually run the motor.

If the motor can be test run, another test to perform is the hot spot test. Motors dissipate a certain amount of heat while operating, heat which should be more or less evenly distributed throughout the frame. Leave the motor running for a while and then touch it in spots to see if there are any areas which are unusually warm. Be careful to avoid burns when doing this. A motor which is burning up will usually smell and smoke from the cooking insulation. Any type of burning is a sign of trouble. It is normal for some freshly baked windings to smell a little on the first run as the winding experiences its operating temperature for the first time, but this should shortly subside.

If a motor is marginally suspected of burning, a thermocouple connected to a pyrometer can be placed on the frame while the motor is running. If the reading on the meter is higher than the designed temperature rise, there may be a ventilation problem. Are the coils spaced enough apart outside the slots, or are they tightly bunched together? Some motors depend on this spacing for convective cooling. If that is the problem, the winding will have to be pried apart the spacers placed between the coils. Otherwise, the whole motor may have to be rewound.

# Chapter 6

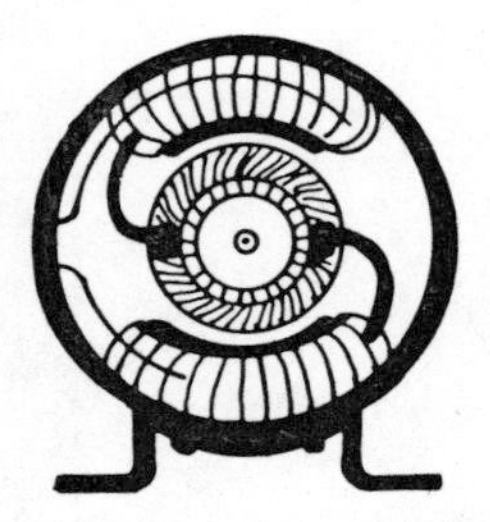

# General Classification of AC Motors

The primary classification of alternating current motors consists of two divisions: Singlephase and polyphase. These names are self explanatory to those who have some knowledge of motor windings. A single-phase motor is one in which the winding consists of a single set of coils, so connected that they generate a single wave of alternating magnetomotive force.

A polyphase motor, on the other hand, has two or more phase windings which are so distributed or connected that two or more waves of alternating magnetomotive force, different in phase, are produced at the same time. Thus a two-phase motor has two phase windings so distributed around the field that the result is two separate magnetomotive force waves which are 90 electrical degrees apart.

Manufactured mostly in the fractional horsepower sizes, the single-phase motor has the widest use of any electric motor. This comes about partly because the majority of them are in service in homes, office and stores where the only source of current supply comes from the lighting circuit. Single-phase motors of various kinds are made in larger sizes to meet special conditions, but their use cannot be said to be general.

## SINGLE-PHASE

Single-phase motors can be classified, according to the principle of operation, into five general types: repulsion, induction, series, capacitor and synchronous. There are subdivisions to some of these types, and in some cases, overlapping features. Each type or characteristic will be considered in order. It should be noted here that single-phase motors of the induction types are not inherently self-starting, thus needing some form of auxiliary winding to bring the rotor up to speed. The two most commonly used methods of starting single-phase induction motors are by split-phase, and by repulsion.

Single-phase straight repulsion motors enjoy a very limited use, and are not to be confused with the more popular repulsion-start, induction-run type, more popularly known as "repulsion-induction." Straight repulsion motors employ a running winding wound on the stator, and a wound rotor of conventional type. There is no electrical connection between the armature winding and the stator winding. Current is induced into the armature winding from the transformer action of the line supplied running winding. The brushes used in a motor of this type are short circuited together, and placed in such a position as to cause the induced current of the armature to create the torque necessary to run the motor. Changing the direction of rotation of these motors is accomplished by moving the brushes to a point either side of the neutral position. See Fig. 6-1.

Single-phase straight repulsion motors have been manufactured which make use of a third winding, and which are known as compensated repulsion motors. In addition to the running winding and the winding on the armature or rotor, there is another winding incorporated in the field ring and which is connected to the commutator of the armature by means of brushes. The compensating winding has no connection to the running winding or to the power supply. The object of the compensating winding is to reduce spark-

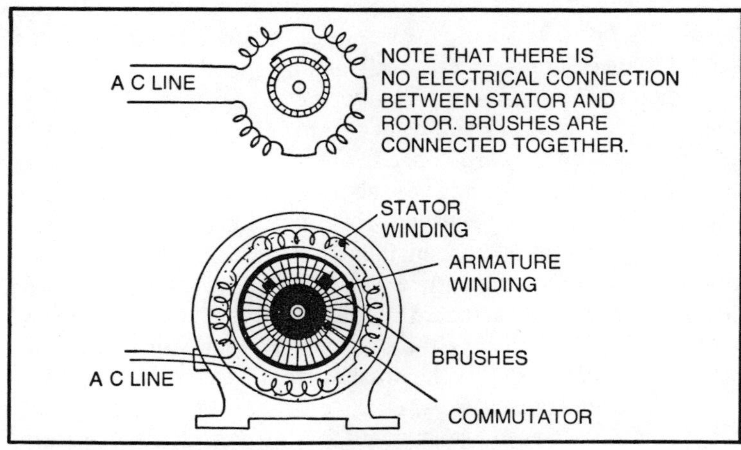


Fig. 6-1. Diagram of straight repulsion motor circuits.

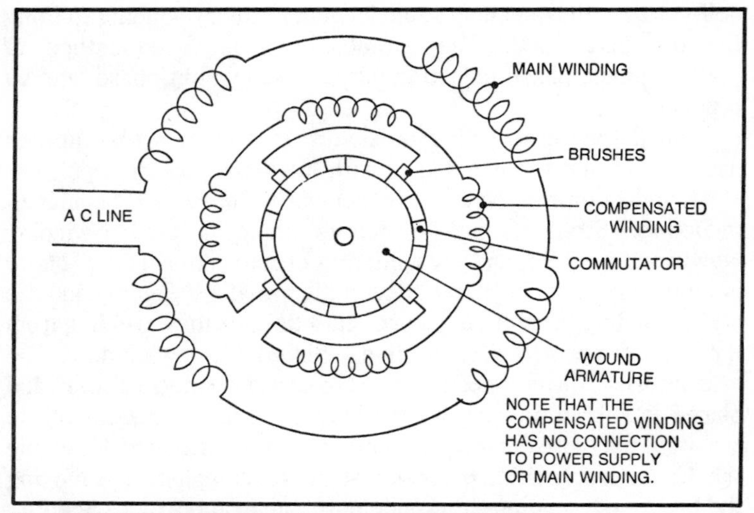

Fig. 6-2. Schematic diagram of a compensated repulsion motor.

ing at the brushes on the one hand and to increase the power factor on the other. A schematic diagram of such a motor is shown in Fig. 6-2.

The repulsion-start, induction-run motor is one of the three leading types of fractional horse power motors in use, and its name comes as a result of its operating characteristics. Like the straight repulsion motor this type has a running winding and a wound rotor. The winding on the rotor, or armature, however in this case, can be considered as an auxiliary winding as it is used in starting only, and its sole function is to supply the torque essential to bring the motor up to the lower limits of a normal running speed.

When starting, a motor of this type operates as straight repulsion motor. As soon as sufficient speed is attained to cause the motor to produce the torque needed to run as a straight induction motor, the armature windings are short-circuited together with the result that at normal speeds the rotor functions the same as the squirrel-cage rotor of a plain induction motor. The method of short-circuiting the armature windings in a repulsion-start, induction-run motor is actuated by centrifugal weights, and on many motors of this general type the same device that shorts the commutator segments also lifts the brushes from the commutator. This not only prevents useless wear on the brushes, but tends to reduce minor sparking caused by incomplete shorting of the armature windings and radio interference. A wiring diagram showing the connections of a repulsion-start, induction-run motor is shown in Fig. 6-3.

The split-phase motor is another popular type, the use of whicn is generally confined to the operation of small machinery where the starting torque is not too high. A motor of this type employs two windings, both wound on the stator, a main running winding, and an auxiliary or starting winding. The rotor is of the solid, squirrel-cage type. A simple centrifugal switch disconnects the starting winding from the line as soon as the speed necessary to operate as a plain induction motor is reached.

The main winding of a split-phase motor is arranged in pole groups as is usually found in other single-phase machines. The starting winding is wound in coils above and *between* the main winding coils, the reason being that these starting coils will set up a rotating field a certain number of slots distant from the effect of the main coils, and cause the rotor to revolve. The starting winding is of no further use as soon as the rotor attains a normal running speed, hence the automatic switch cuts it from the line, conserving current and preventing the high resistance starting winding from over heating and burning out. Figure 6-4 shows the circuits of a motor for this type.

Single-phase alternating-current motors of the series types are more often called universal motors, in as much as most of them are built to operate on ac or dc current of like voltage. In general these motors are of miniature size and find a rather wide use in fans,

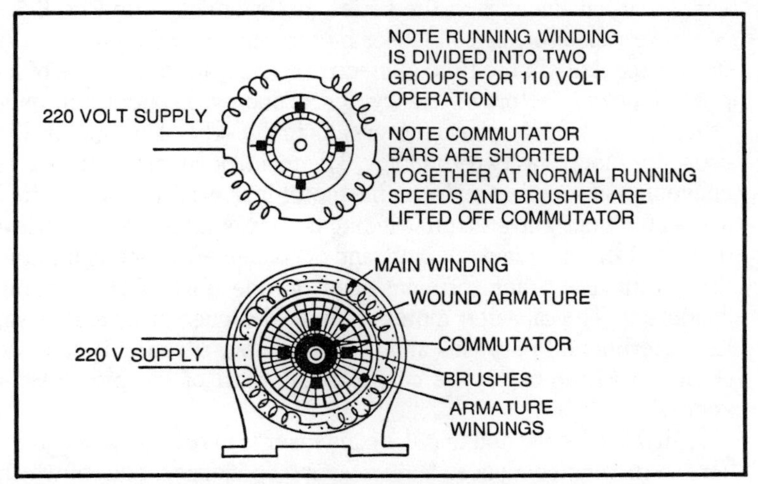


Fig. 6-3. Circuits of a repulsion-start, induction-run motor.

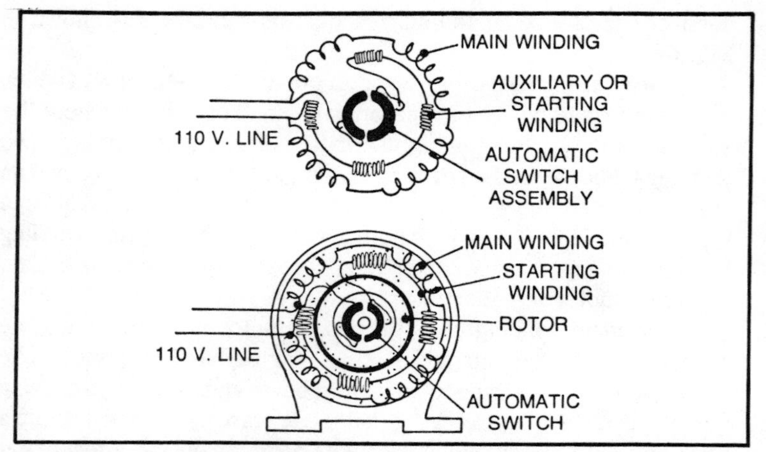


Fig. 6-4. Circuit of a split-phase alternating current motor.

carpet sweepers, drink mixers, juice extractors, sewing machine motors, and for similar tasks where the duty is intermittent and first cost an important consideration.

Series motors have a field—usually two pole—that is wound like the coils in a straight dc machine. The armature is wound like a dc armature and is in series with the field. Motors of this kind are finding less favor than formerly due to the fact that they cause more or less interference with radio reception. The judicious use of bypass condensers will, however, quiet most of these motors. The simple circuit diagram of the series motor is shown in Fig. 6-5.

The capacitor, or condenser motor, has moved rapidly to the front, especially in the field of electric refrigeration, because of its greater power factor, efficiency and economy of operation, over other types of single-phase motors. As in the split-phase motor, the capacitor motor has two windings. The main, or running winding, is conventional in design and with the motor in operation, is connected across the line. The starting winding is approximately 90 electrical degrees from the main winding, and is connected at one end to the line, while the other terminal ends on one post of the running condenser. The capacitor motor has been designed in several ways, for experimental purposes and otherwise, but the circuit diagram shown in Fig. 6-6 may be considered typical of the principle in general use.

In this case we find that the condenser is in reality two condensers, a running condenser and a starting condenser, both of which are incorporated in a metal container outside of the motor. The

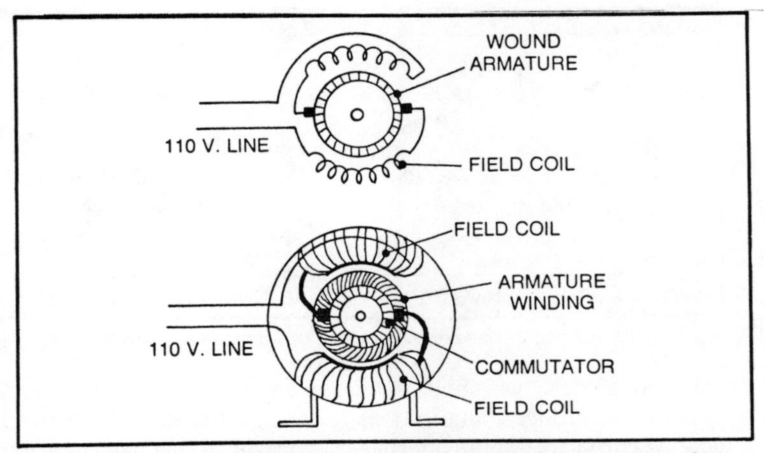


Fig. 6-5. Wiring diagram of a single-phase motor of the series or universal type.

starting condenser, as the name implies, is brought into use while the motor is starting, and is connected on one side to the line, and on the other to the starting winding. A centrifugal switch on the rotor disconnects the starting condenser as soon as the proper speed is obtained. Capacitor motors cause no interference with radio reception.

The synchronous alternating current motor has a use in certain industrial fields when precise speed regulation is necessary, or when energy conservation is desirable. Because the induction motor has a low power factor at light loads, substituting a synchronous motor for the same application will result in more efficient

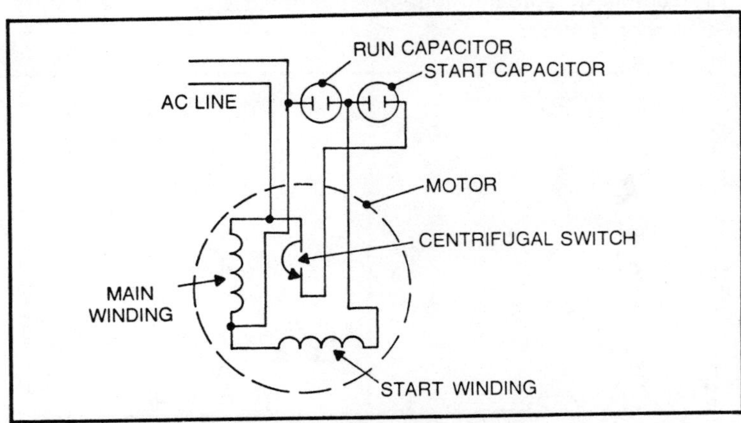


Fig. 6-6. Theoretical diagram of one form of capacitor motor.

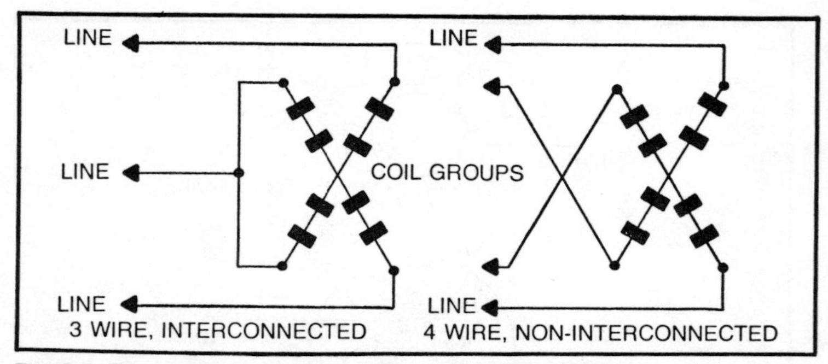


Fig. 6-7. Two phase three wire motors and two phase four wire motors.

operation. The synchronous motor, however, is almost never used at home for ordinary applications (and no, those clock motors are not synchronous, but hysterisis motors or reluctance motors). Aside from the large industrial centers few service men will even be called upon to repair motors of this type.

## POLYPHASE

Polyphase motors comprise the second group of alternating current motors. These may be two or three phase, the term polyphase simply refers to all motors with more than one phase. These motors have their windings connected and placed so as to produce a rotating magnetic flux when polyphase current is supplied. The windings are placed in the stator, while the rotor, which in reality is the "field" of the machine, is not connected to any source of electrical energy. The rotor is basically a solid piece of metal, which is extremely durable and long lasting, hence the great reliability of induction motors.

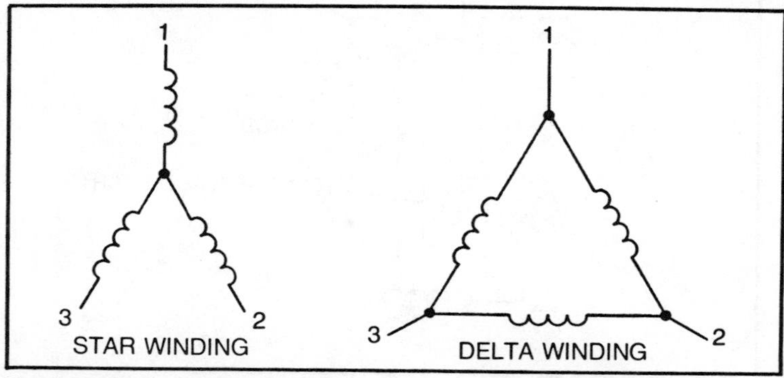


Fig. 6-8. Three phase motor circuits.

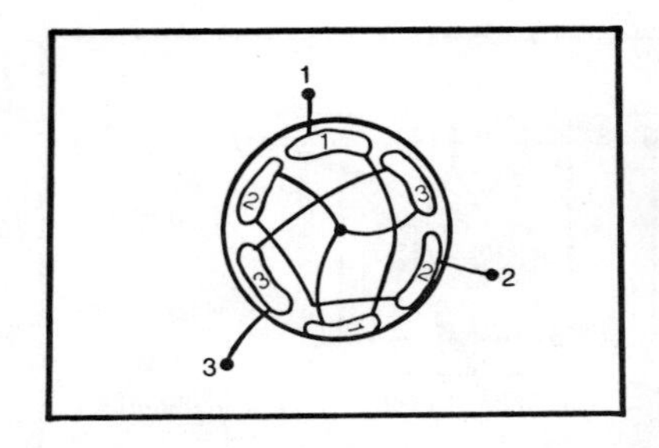

Fig. 6-9. Coil layout inside a 3 phase, 2 pole, star motor.

Current is induced in the rotor as the flux lines revolving about the primary cut the conductors of the secondary. Polyphase motors produce between two to four times their full load torque when starting, so the motors are capable of accelerating up to full speed without the need of any auxiliary windings or starting devices, as in the case of single phase motors. The torque then begins to rise as the motor accelerates, reaching a peak at perhaps 75% of synchronous speed, where then it falls to normal full load torque at the running speed.

Two phase motors are almost never seen, because two phase power is found in only a few remote locations. There are basically two types of these motors, however, so named because of the number of leads each motor has. A two phase, 4 wire motor has two windings which are not connected internally. Two phase, 3 wire motors do have interconnected windings. Figure 6-7 shows both types. It is possible to run a 4 wire motor off a 3 wire line, if two of the leads are joined together. The motor is reversible by interchanging either pair of the two supply leads of one phase.

The number of coil groups in Fig. 6-7 has to do with the number of poles which the motor has and also to some extent on the voltage at which the motor operates. Sometimes, the windings may be arranged so that groups of coils may be connected in series or

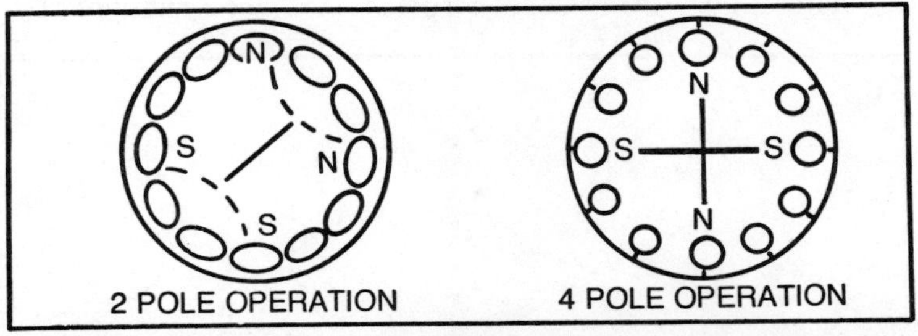

Fig. 6-10. Left, connecting the windings in star gives 6 poles, while at right, connecting them in delta gives 12 poles.

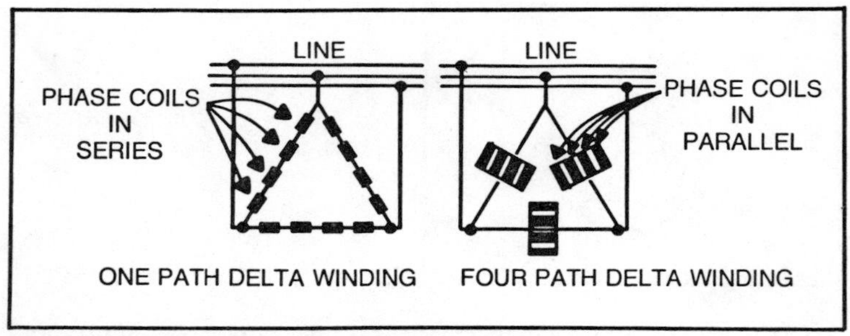


Fig. 6-11. A single path and a four path delta winding.

parallel, the former to run at a high voltage, and the latter, for a lower voltage.

Three phase motors are far more common than two phase types. These also have their windings arranged in different manners depending on the interconnection of phases, number of poles, number of paths, and other factors. Three phase motors are classified into two types based on the winding connections: a three phase star or Y, and a three phase delta. Figure 6-8 shows both types.

In Fig. 6-8 it can be seen that in the star winding, each phase is connected at a central point, while in the delta winding, each end of the phase windings are connected between two of the supply leads. These phase windings, be it either a star or delta winding, are not just one coil, but rather several, depending on the number of poles and the voltage. Figure 6-9 shows how a two pole star primary would be laid out inside a motor frame. Notice that there are six groups inside the motor, two for each phase. Thus, when a three phase motor is said to be of a two or four pole design, it does not mean that there are that many poles, or coil groups inside the motor, as in a single phase winding. Rather, the total number of poles inside the motor is equal to the number of poles of the motor multiplied by

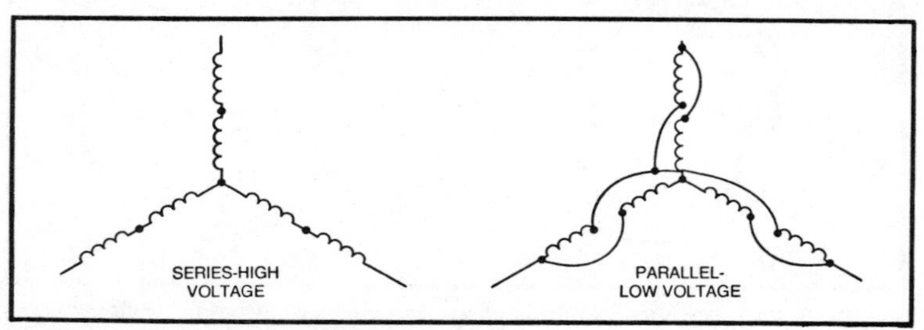


Fig. 6-12. Dual voltage connections.

the number of phases. In Fig. 6-9, 2 × 3 = 6. If there were 12 groups inside the motor, the motor would have 4 poles (4 × 3 = 12), and so on.

Figure 6-10 shows the star and delta again, this time with four phase coils in each phase, or leg. These coil groups are not laced consecutively in the stator, rather in succession. In practice a coil belonging to one phase group is followed by a coil of the second phase group, and that in turn by a coil of the third, and so on around the core. Figure 6-9 shows this pattern nicely.

Most motors are designed to be of one type only, either the star or delta. Consider this winding with 12 poles, or coil groups, in it. If the motor leads are designed so that they may be connected in *either* a star or delta, the number of actual poles produced, and hence the speed of the motor, will change.

If the six leads are connected in a star, the motor will run at a higher speed because two coil groups will have the same polarity and act as if they are only one pole, while connecting them in delta allows the poles to remain discrete. This does not mean that star motors have higher synchronous speeds than delta motors. Only in this switchable motor is that true. By arranging the phase coils in different orders, one is able to attain any configuration or speed with either winding.

Term "1 path", "2 path", "4 path", etc., refers to the number of current paths through each of the phase windings. In Fig. 6-11 there are two delta primaries. In each, there are four phase coils per phase, but in one, they are connected in series, while in the other, they are connected in parallel. This technique is often used to set the motor for the correct voltage, as series or parallel windings differ in how much current and voltage they require.

Figure 6-12 shows how a star winding may be connected in two ways to operate at two different voltages. Keep in mind that these coils may have any number of groups depending on how many poles the motor has. Figure 6-13 shows how a six lead motor may be

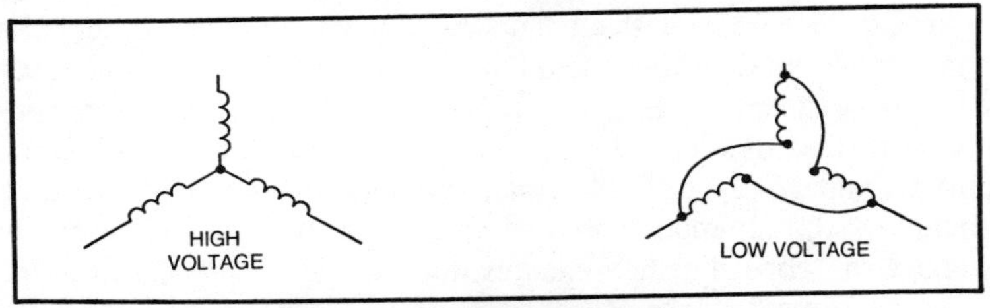


Fig. 6-13. Dual voltage by switching from star to delta.

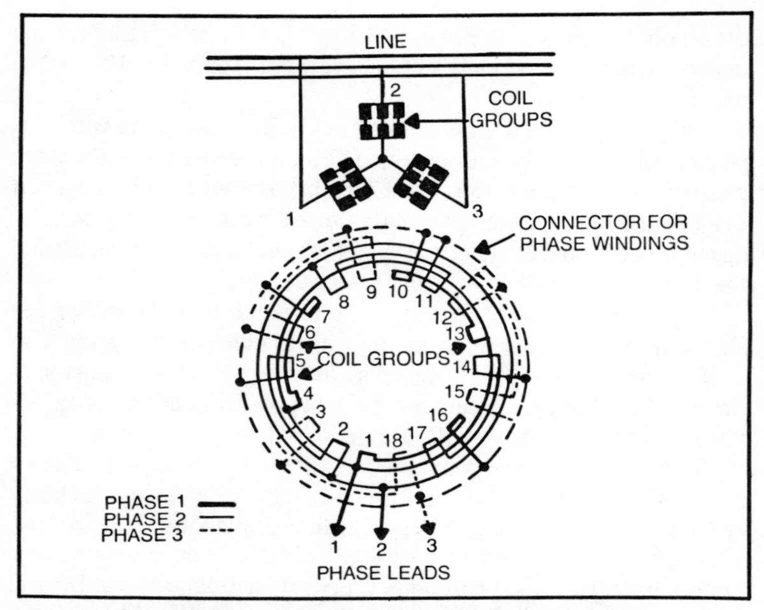


**Fig. 6-14. Circuit diagram of a 3-phase, 6 pole, 3 path, star winding.**

connected in star or delta to give dual voltage operation. This does not change the speed of the motor, as in Fig. 6-10, because the direction in which the coils are wound and the way the current is fed into them keeps the number of poles constant. All of the connections for different speeds, voltage, etc., are given on the motor name-plate, so one does not have to memorize all possible combinations, but an understanding of how the principles apply is helpful and in some cases necessary.

Figure 6-14 shows the layout of a 3 phase, 6 pole, 3 path, star winding. Such a motor has 18 internal poles (3 × 6), and may have a synchronous speed of 1200 rpm (if the supply frequency is 60 hz). At first this may seem like a great deal of confusion, but a little studying will reveal the structure. Phase two, for example, feeds three coils in parallel, these then are connected to three other coils in series, and then these join together at the common. Looking at Fig. 6-14 shows which coils they are. From the #2 line, the connections go to coils 2, 8, and 17. Coil 2 is then connected to coil 5, coil 8 to 11, and 17 to 14. These coils, numbers 5, 11, and 14, are then joined at the common. Note that phase 2 has 6 equally spaced coils about the core. Further examination will show the patterns for phases 1 and 3.

# Chapter 7

# Split-Loop Armature Winding

The several types of split-loop windings have certain advantages over the regular plain loop winding. Some of the advantages of the split-loop winding are that it aids in giving better space distribution across the ends of the winding, and that it gives a more even mechanical and electrical balance. The latter is especially true in the case of the parallel wound type of split-loop winding. See Fig. 7-1.

Here we find that the coils are wound in pairs, parallel to each other and one on each side of the shaft. Consequently each pair of coils that are parallel to each other are of the same length, and hence will be of the same weight and electrical resistance. In the plain loop winding each one of the coils will be slightly different in these respects while in the split-parallel-loop winding the number of unlike coils is reduced by one half. Another advantage of this type of winding is that it can be wound with more than one wire in hand if required. The only valid objections to this type of winding can be summed up by saying that it is slightly harder to connect up, that it is easier to make mistakes than in the plain loop winding, and that the wires must be cut at the completion of each coil.

## CHORDED WINDINGS

The different types of windings discussed here are all chorded. By "chorded" is meant that the coils are wound less than full pitch. If an armature has, for instance, 14 slots, the full pitch for a two pole machine would be in slots 1 and 8, those diametrically opposite each other. When the windings are placed less than full pitch, as in slots 1 and 6, or 1 and 7, the winding spans less than the full distance between pole centers and is called a chorded winding.

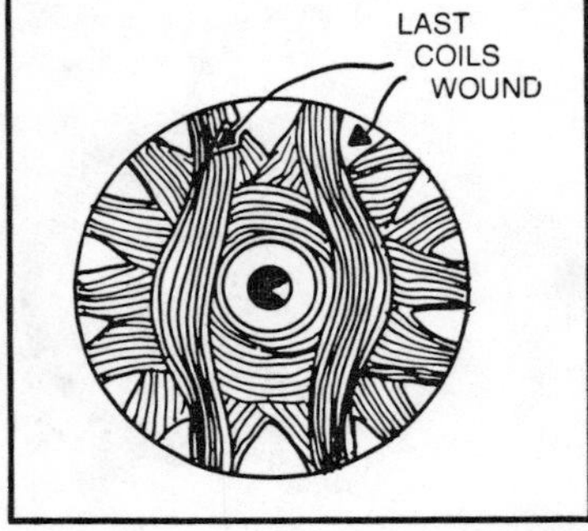


Fig. 7-1. End view of a parallel-split loop winding.

Figure 7-2 shows the method of starting a parallel split-loop winding. Arbitrarily, an armature having 14 slots has been chosen for purposes of explanation, but any armature with an even number of slots can be wound in the same general manner. The first coil is started in the slot selected as number 1 and comes back in slot 7, then back through slot 1 and so on around until the correct number of turns have been placed. The wire is now cut at the commutator end, leaving ample length to reach the proper commutator bar with an inch or two left over. A piece of white sleeving should now be slipped over the starting end of the coil while a piece of red sleeving is placed around the finishing end. This method of marking is followed on each coil as it is completed so that all starting ends and all finishing ends will be color coded the same.

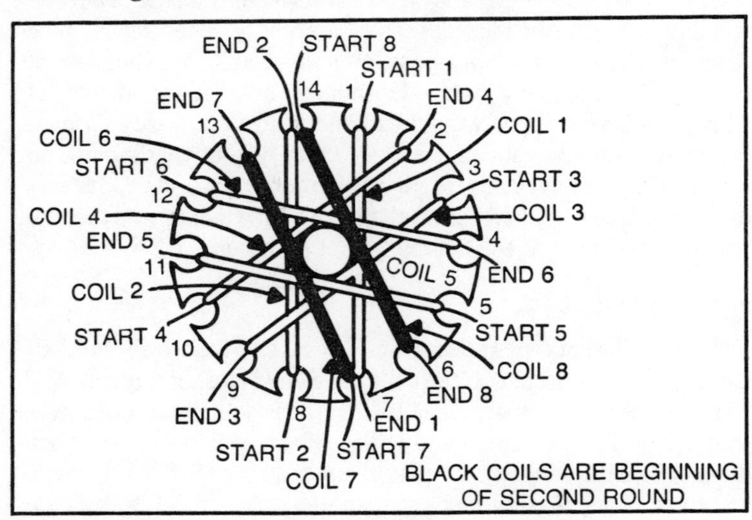


Fig. 7-2. Starting a parallel-split loop winding.

When the first coil is completed the armature is turned one half way around so as to be in position to receive the second coil. An inspection of Fig. 7-2 will show that the second coil goes into slot 8 and 14. A white and a red sleeve are slipped over the starting and the finishing ends of the second coil. The armature is now again turned until slot 3 is in the top position. The third coil is found in slots 3 and 9, and the parallel coil in slots 10 and 2, starting in slot 10 and finishing in slot 2. The fifth coil is started in slot 5 and finishes in slot 11, while its companion coil, No. 6, starts in slot 12 and ends in slot 4.

We have now arrived at a point where all the slots hold one coil side except slots 6 and 13 which are still vacant. These slots are opposite each other and can not hold sides of the same coil, hence it is at this point where we must start to overlap and start the second layer. Coil 7 starts in slot 7 on top of the first coil laid (No. 1) and ends in the otherwise vacant slot No. 13. Moving over to slot 14 we start the 8th coil in top of coil No. 2 and finish it in slot 6 which was heretofore vacant. The rest of the coils forming the top layer are now wound in the following order: Coil 9 starts in slot 9 and finishes in slot 1, coil 10 starts in slot 2 and finishes in slot 8, coil 11 starts in slot 11 and finishes in slot 3, coil 12 starts in slot 4 and finishes in slot 10, coil 13 starts in slot 13 and finishes in slot 5, coil 14 starts in slot 6 and ends in slot 12. See Fig. 7-3.

In the type of winding under consideration at this point there is a slight difference in the arrangement of the coils depending upon

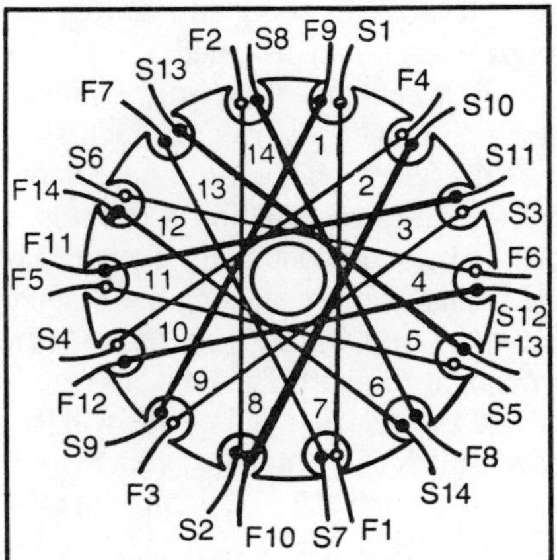

COMPLETE DIAGRAM FOR A
14 SLOT ARMATURE
HEAVY LINES DENOTE TOP COILS

Fig. 7-3. Complete diagram of a parallel-split loop winding.

the fact of whether or not the number of armature slots is divisible by 2 or 4. Rather than take up time and space by going into details of the two variations it has been thought better to condense such information into a table giving the winding data for armatures having from 10 to 24 slots. This winding chart is shown in Table 7-1.

Another form of the chorded split-loop winding, employed mainly on small two pole motor armatures, is what might be called a divided split-pitch loop winding. This form of winding differs from the one just considered in that two full coils are wound in but three slots. As will be seen in the typical example of this type of winding shown in Fig. 7-4, two coil sides are wound in slot No. 1 with the remaining sides of each coil going to adjacent slots on the same side of the shaft. In this case we have an armature with 24 slots. The full pitch for this armature would be 12 slots or 1 and 13, a winding which would tend to "pile up" at the ends because each coil would have to cross the largest number of other coils besides bending around the shaft.

By making the winding short pitch, or chorded, we eliminate a great part of this trouble. If we make the pitch 1 and 11 instead of 1 and 13 we are covering two slots less than full pitch, which helps some but may not solve all of the difficulty of getting the windings in the space allowed. As a further aid to securing a better distribution of the coil ends the divided, or split-pitch winding is often used. This winding is wound with one wire in hand, and can be used on armatures having two or four times as many commutator bars as slots. It can be loop wound with a continuous wire, in which case the loops must be color coded alternately.

In winding the armature shown in Fig. 7-4 the first coil is started in slot 1, carried through slot 11, and back through slot 1 until the full number of turns have been wound. The finish of the first coil is in the same position as the beginning—at the commutator end of slot 1—and a loop long enough to reach the proper commutator bar is made before starting the second coil. Coil two is also started (from the loop) in slot 1, but instead of also going through slot 11 it is wound into slot 10. Thus the pitch of the coils is split between 1 and 10 and 1 and 9. Half of the coils will have a pitch of 10 slots, half will have a pitch of 9 slots, the average for the whole winding will be 9½ slots, or a pitch of 9.5. Coil three starts in slot 2, carries through slot 12 and finishes with a loop in slot 2. Coil four starts in slot 2, winds in slot 11 and finishes with a loop in slot 2 ready to commence coil five in slot 3. This method is carried on around the armature until each slot contains four coil sides.

## SPLIT V LOOP

Still another type of loop winding is known as the split V loop. The chief advantage of this form of winding will be found where there is little space between the bottoms of the slots and the shaft. Because the coils are split, one coil end from each V shaped pair goes on each side of the shaft as seen in Fig. 7-5. While this winding is wound with loops between coils it can not be connected like a regular loop winding. In this case the loops must be cut before connecting to the commutator and it is important that the colored sleeving be around each individual starting and finishing lead and not merely slipped over the loop. Then when the loops are cut the proper designation can still be told. To accomplish this a dozen or more pieces of sleeving, alternately red and white, are slipped over the wire before starting the winding.

A white sleeve can be left on each starting lead and a red one on the finishing end. A loop is then made in the wire, then another white sleeve is put in place, and another coil is wound. Whenever the supply of sleeves on the wire runs out the wire must be cut and a new supply slipped over the wire as in the beginning. With this kind

**Table 7-1. Winding Chart for Split Loop Armatures Having From 10 to 24 Slots. Compiled From Tables In Rewinding Small Motors, By Braymer and Roe.**

| | 10 SLOTS | | 12 SLOTS | | 14 SLOTS | | 16 SLOTS | | 18 SLOTS | | 20 SLOTS | | 22 SLOTS | | 24 SLOTS | |
|---|---|---|---|---|---|---|---|---|---|---|---|---|---|---|---|---|
| Coil number | Start in slot No. | Finish in slot No. | Start in slot No. | Finish in slot No. | Start in slot No. | Finish in slot No. | Start in slot No. | Finish in slot No. | Start in slot No. | Finish in slot No. | Start in slot No. | Finish in slot No. | Start in slot No. | Finish in slot No. | Start in slot No. | Finish in slot No. |
| 1 | 1 | 5 | 1 | 6 | 1 | 7 | 1 | 8 | 1 | 9 | 1 | 10 | 1 | 11 | 1 | 12 |
| 2 | 6 | 10 | 7 | 12 | 8 | 14 | 9 | 16 | 10 | 18 | 11 | 20 | 12 | 22 | 13 | 24 |
| 3 | 3 | 7 | 3 | 8 | 3 | 9 | 3 | 10 | 3 | 11 | 3 | 12 | 3 | 13 | 3 | 14 |
| 4 | 8 | 2 | 9 | 2 | 10 | 2 | 11 | 2 | 12 | 2 | 13 | 2 | 14 | 2 | 15 | 2 |
| 5 | 5 | 9 | 5 | 10 | 5 | 11 | 5 | 12 | 5 | 13 | 5 | 14 | 5 | 15 | 5 | 16 |
| 6 | 10 | 4 | 11 | 4 | 12 | 4 | 13 | 4 | 14 | 4 | 15 | 4 | 16 | 4 | 17 | 4 |
| 7 | 7 | 1 | 6 | 11 | 7 | 13 | 7 | 14 | 7 | 15 | 7 | 16 | 7 | 17 | 7 | 18 |
| 8 | 8 | 6 | 12 | 5 | 14 | 6 | 15 | 6 | 16 | 6 | 17 | 6 | 18 | 6 | 19 | 6 |
| 9 | 9 | 3 | 9 | 1 | 9 | 1 | 8 | 15 | 9 | 17 | 9 | 18 | 9 | 19 | 9 | 20 |
| 10 | 4 | 5 | 8 | 7 | 8 | 6 | 16 | 7 | 16 | 6 | 19 | 8 | 20 | 8 | 21 | 8 |
| 11 | | | 10 | 3 | 11 | 5 | 10 | 1 | 11 | 1 | 10 | 19 | 11 | 21 | 11 | 22 |
| 12 | | | 4 | 9 | 4 | 10 | 2 | 9 | 2 | 10 | 20 | 9 | 22 | 10 | 22 | 10 |
| 13 | | | | | 15 | 3 | 12 | 3 | 13 | 3 | 12 | 1 | 13 | 23 | 12 | 25 |
| 14 | | | | | 6 | 15 | 4 | 11 | 4 | 12 | 2 | 11 | 2 | 12 | 24 | 11 |
| 15 | | | | | | | 14 | 5 | 15 | 5 | 14 | 3 | 15 | 3 | 14 | 1 |
| 16 | | | | | | | 6 | 13 | 6 | 14 | 4 | 15 | 4 | 14 | 2 | 15 |
| 17 | | | | | | | | | 17 | 7 | 16 | 5 | 17 | 5 | 16 | 3 |
| 18 | | | | | | | | | 6 | 16 | 6 | 15 | 6 | 16 | 4 | 15 |
| 19 | | | | | | | | | | | 16 | 7 | 19 | 7 | 16 | 5 |
| 20 | | | | | | | | | | | 6 | 17 | 8 | 16 | 6 | 17 |
| 21 | | | | | | | | | | | | | 21 | 9 | 20 | 7 |
| 22 | | | | | | | | | | | | | 10 | 20 | 6 | 19 |
| 23 | | | | | | | | | | | | | | | 22 | 9 |
| 24 | | | | | | | | | | | | | | | 10 | 21 |

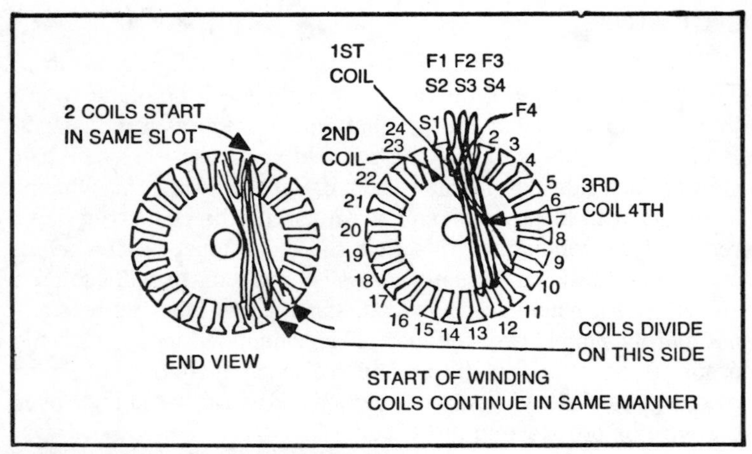


Fig. 7-4. Divided split-pitch loop winding.

of winding a slot will be completed as soon as the two coils are wound, and it can be wedged at once if desired. This winding can only be used on an even number of slots.

Another form of winding which is adaptable particularly for low voltage armatures where the gauge of the wire is large and the turns few, is what is known as a diametrically split winding. As a winding of this kind must be full pitch to be symmetrical, it can only be used where the slots count up to an even number.

In the diametrically split winding each coil is divided in half as it is wound, the two halves going on opposite sides of the shaft. When the winding is started several turns are passed to the right or left of the shaft, then a like number are wound on the other side of the shaft, and so on back and forth until the full number of turns have been made. Figure 7-6 will give a clear idea of the method of placing a winding of this type.

In the foregoing paragraphs an effort has been made to give a general idea of the various types of windings that are often used in small motors and generators operating on direct current. More detailed treatment would take up an undue amount of space. It is seldom that a stripped armature comes to the rewinding shop, but when it does a general knowledge of the different windings may be of value during the process of "engineering" a winding to replace the original.

Usually, and also fortunately for the rewinder, the armature is usually delivered for repairs with the old winding more or less intact, and the necessary data for winding can be secured before stripping. If it is known or can be ascertained that the now defective winding has given long and satisfactory service the rewinder can hardly do better than to copy the old winding in such details as pitch, size of wire, kind of insulation, number of turns, etc. Naturally many rewinders will have pet methods that have proved their worth, new and better materials may be on the market, and it is quite all right to make legitimate substitutions that are known to be good.

When an armature winding is an unknown quantity it should be hand stripped down to a point where the winder will be absolutely sure that he is familiar with the commutator connections and all other information. In such a case, or where this information is already on file from previous jobs of the same kind, the balance of the stripping job can be done in the quickest and best manner.

One method of stripping small armatures that has found favor in many a shop is to cut off all the coil ends at one end of the core. This can be done with a hacksaw or on a lathe. The armature is then placed over a low burning gas fire or in a hot oven and allowed to "bake". The commutator can be removed or not, depending on the proximity to the windings, its construction and the type of heat used to bake the old winding. Commutators should never be placed over a flame.

After the winding has been thoroughly baked the shaft can be bumped sharply when in a vertical position and the winding will slip out of the core. If the baking has been thorough the slot insulation will usually come out with the coils, but if not it can easily be cleaned out with an old hacksaw blade.

Usually armatures are wound with the coil ends at the drive end closely hugging the shaft. This allows more room for distributing the coil ends and shortens the distance the wires must travel

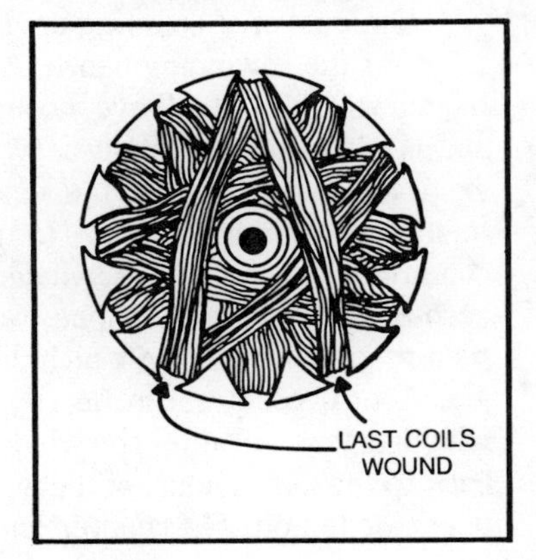


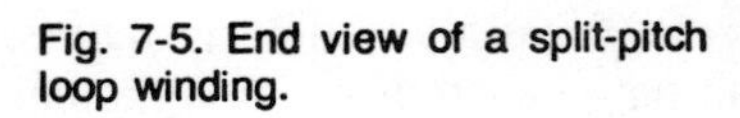
Fig. 7-5. End view of a split-pitch loop winding.

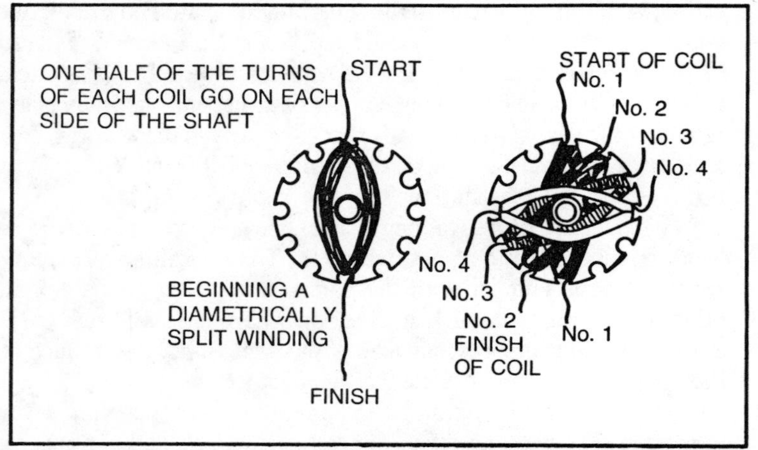


Fig. 7-6. A diametrically split winding. This is wound full pitch or "on the half."

between slots. Some armatures, however, are wound with a hollow space between the shaft and the windings. The reason for this is either because there is a projection of the bearing boss of the end bell, or else to improve ventilation of the winding.

To form a hollow end in the winding a wooden spool is first slipped over the shaft, and the windings are then wound tight around the spool. Every shop should have a set of these spools available in sizes to fit shafts from ½ inch up to at least 1½ inches in diameter. An assortment of outside diameters will also be necessary in as much as the spool must be smaller than the radius of the bottom of the armature slots. Usually the spool will be at least a ¼ inch below the bottom of the slots. After the winding is completed the spool can easily be slipped off of the shaft by giving a quarter turn.

Some form of counting device is a very handy piece of equipment for the rewinding bench. Armatures and stators used on the higher voltages often have a considerable number of turns per coil and an exact count of the turns wound is very desirable. In the large shops and in the factories the winder will be given a job and allowed to stay with it until through. In the small shop, however, different conditions prevail. The rewinder may have a dozen other duties, such as answering the phone, waiting on customers, doing minor rush repairs, etc., all of which distract attention from the winding job. A turn counter can be rigged up for use when laying hand windings that will allow the winder to stop at any time and yet come back to the job an hour, or a day, or a week later and know exactly where he left off. The counter can be operated by the hand or foot in some ingenious manner. The writer uses one made from the trip mileage assembly from an old speedometer which is very satisfactory. Counters for all kinds of industrial jobs can be purchased at moderate prices, but if one is purchased get one that can be reset to zero quickly.

# Chapter 8

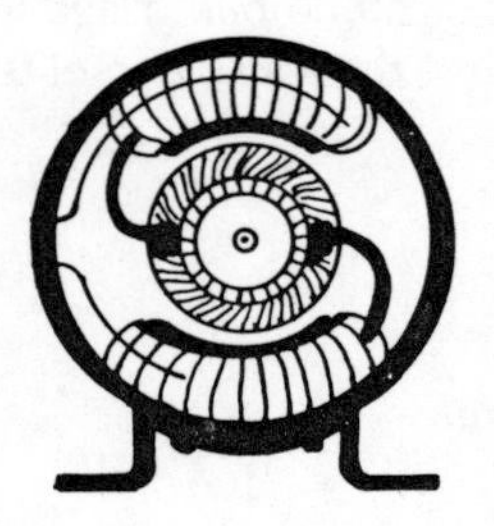

# Loop Windings for Small Armatures

In the course of a year many different types of armatures will come into the rewinding shop and it is necessary for the winder to have a comprehensive knowledge of the various forms of windings. One of the most common types is known as the loop winding. On a loop wound armature one complete coil end, or loop, can be seen on the driving end. This is, of course, the last coil wound and is the only one not partially covered by adjacent coil ends. See Fig. 8-1.

Because consecutive coils finish and start in slots next to each other, and because both of these leads — the finish of one coil and the start of the next—go to the same bar on the commutator, it is not only possible but often practicable to wind the armature with a continuous wire. When this is done a loop is made long enough to reach the proper commutator bar at the finish of one coil, and the wire is doubled back to start the next coil. A study of Fig. 8-2 will show the manner in which this is accomplished.

Loop windings are quite popular on small armatures, such as are used in vacuum cleaner, fan and most universal type fractional horsepower motors. One of the reasons for this is that such armatures are most often wound on armature winding machines, and the loop winding is most adaptable to this form of production. Most rewinders find it not only tedious but unsatisfactory to attempt to rewind very small armatures without the aid of a winding machine. Most of these armatures are wound with plain enameled wire of small diameter and require an even tension to prevent breaking of the insulating film, resulting in shorts.

Few rewinding shops depending on local trade will find it worthwhile to install an armature winding machine for the comparatively few armatures that can not readily be hand wound. There are many concerns making a specialty of the machine winding of small armatures, and usually an exchange plan is offered giving quick service and low cost. In general, hand winding will be found satisfactory on armatures employing cotton covered wire if the price received for the finished job is in proportion to the labor involved.

In brief, the advantages of the loop winding are that such a winding is quickly wound, and the connections to the commutator are easily made with the smallest chance of error. The chief disadvantages of this form of winding lies in the unequal amount of copper in the various coils, and in unequal mechanical balance. Because the coil ends must be built up, one upon the other, it will be seen that the last coils laid will contain a greater length of wire than will the first coils placed in the slots. Thus there will be a variation of the electrical resistance of the different coils as well as a difference in weight.

Many beginning rewinders experience great difficulty in securing a symmetrical winding because of the tendency of the coil ends to "pile up," with the result that the finished winding occupies too much space, and may even result in preventing its assembly into the machine. There are three tricks well known to experienced rewinders which will overcome this tendency.

The first and probably most important rule to remember is that the *first three coils laid* determine the shape of the finished job, for it is over this position that the last coils overlap. Therefore, particular attention must be given to the first few coils in the way of pressing them neatly and closely against the core and close to the shaft.

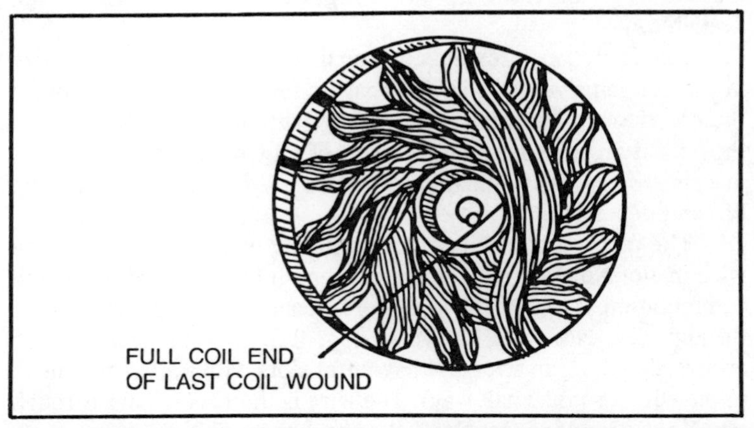


Fig. 8-1. Appearance of the end of loop winding. Only one full coil end can be seen.

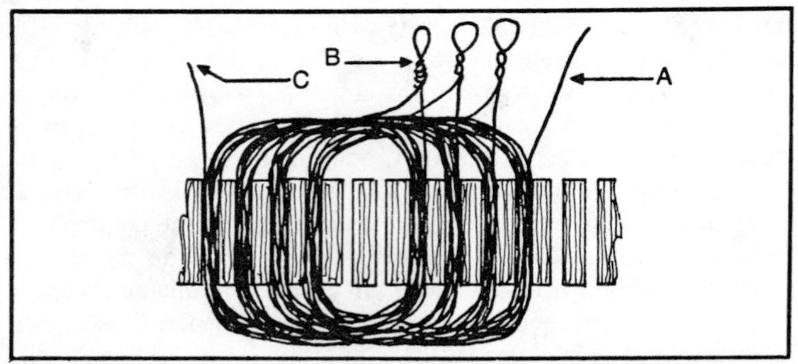


Fig. 8-2. How a loop winding would appear if wound on a flat surface. A, beginning end of wire. B, loops left between coils so as to have continuous wire. C, wire to reel in used winding.

## FIBRE END LAMINATIONS

In this connection it might be well to state that fibre end laminations should be used whenever possible. These fibre washers can be purchased in sizes and shapes to fit all standard type cores, and a reasonable stock of them should be maintained. Their use eliminates one of the greatest troubles that bothers the winder, that of grounds at the corners where the wires leave the slots. When no factory cut end washers are immediately available, some rewinders take the time to hand cut a set from sheet fibre. When no other pattern is at hand one cut a set from sheet fibre. When no other core and used as a pattern, after which it can be replaced. Figure 8-3 shows the use of these fibre laminations, while Fig. 8-4 shows the method of laying coil ends when regular core insulators are not available.

One fairly large rewinding shop discontinued the use of these insulators some years ago with apparent success. However, most of their work was in the field of low voltage automotive armatures, and they used an unusually heavy type of slot paper with generous margins. Fibre end laminations must be used if the utmost in long life and dependability are expected.

The method of placing a loop winding is as follows: Start the wire in slot No. 1, leaving sufficient length to more than reach the commutator bar to which it will later be connected, and bring around through slot No. 7 as shown in Fig. 8-5. (In this case we are arbitrarily taking an armature having 14 slots with a pitch of 1 and 7, or one slot less than half way). The wire is then taken back through slot No. 1, around to slot No. 7, then back through No. 1 and so forth until the required number of turns have been made.

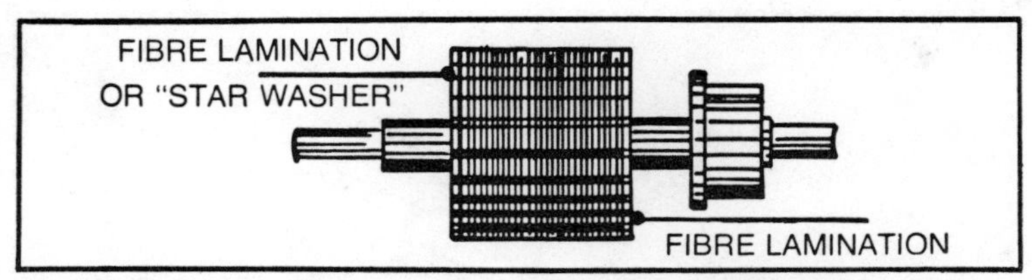

Fig. 8-3. Showing proper use of fibre end laminations pressed on shaft.

The wire is then carried out past the commutator, twisted into a loop, and is brought back into slot No. 2. The coil occupying slots No. 2 and No. 8 is now wound, leaving another loop at its completion. The coil in slots 3 and 9 is now placed, followed by the other eleven coils. At the conclusion of the winding operation there will be 13 looped ends plus the single starting and finishing ends. These two single ends can be twisted together, after which the loops can be placed in the proper commutator bars in order of rotation. As this particular armature is one wire in hand there will be 14 bars, or one bar per loop.

Another rule that the experienced winder follows to avoid bulkiness in the winding is to "fan out" the coil ends. This means that instead of laying the wires in a more or less round bundle as they are in a slot, the winder spreads the wires out fanwise at the ends where they must pass across other coils. This procedure utilizes the space to the best advantage and tends to keep the layers thin and closer to the core. See Fig. 8-6.

Still another method of keeping the coil ends closely packed and symmetrical is to make frequent use of a hammer or small mallet. The blows should not be applied directly to the coil ends but through the medium of a small block of soft wood, one with rounder corners. The blows should not be heavy, as considerable caution

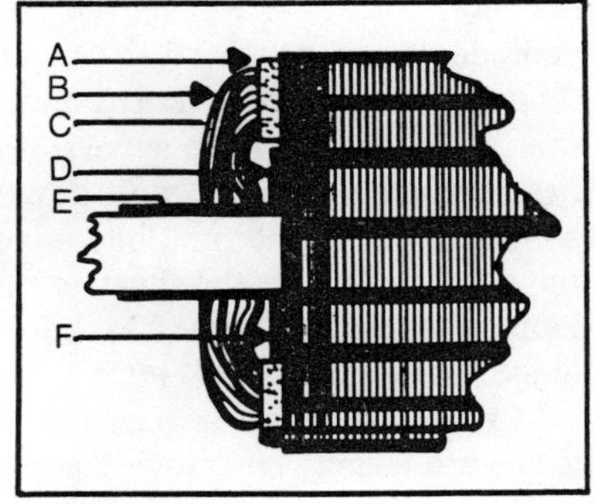

Fig. 8-4. How to insulate armature coils when no fibre laminations are used. A, wedge. B, slot fibre. C, coil end. D, round fibre washer. E, shaft insulation. F, air space.

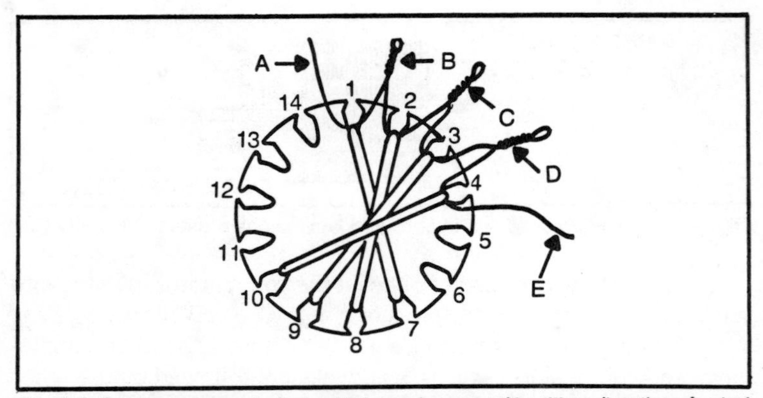


Fig. 8-5. Starting a loop winding, which can be wound in either direction. A, start of first coil. B, end of first coil and beginning of second coil; C, end of second coil, beginning of third; D, end of third coil, beginning of fourth; E, end of fourth coil ready to loop and proceed with fifth coil.

must be used to avoid cutting through the insulation. A little practice will soon show where (and how much) pressure can be applied for the best results.

So far we have considered only the loop winding as wound with a single wire. They can, however, be wound with two or three wires in hand, or else the operator if he wishes can wind two, three or more coils in the same slot with a single wire. In this latter case—using but a single wire—the first coil is wound in the slots, then a loop is formed the same as for the single coil winding, another coil is wound in the same slots and on top of the first one, another loop is formed, and additional coils are wound in the same manner.

In a winding of this kind the number of coils per slot is determined by the relation of commutator bars to slots. Thus if there are 14 slots and 42 bars it will be a case of 42 divided by 14, or 3 coils per slot. With the one wire in hand method of forming more than one coil per slot, each loop will be the start of one coil and the finish of another. However, there is danger of getting the loops from the different coils mixed when assembling to the commutator and for this reason winders use several ways of marking the loops of the first, second, third, etc., coils. Either colored woven sleeving can be slipped over the different loops—such as red for the first, black for the second, blue for the third, or else the winder can leave a short loop on the end of the first coil, a longer one on the second, a still longer one on the third, etc.

When winding an armature with more than one wire in hand the wires are usually cut at the completion of each set of coils and it becomes necessary to brand the ends. This is most often done with colored sleeving but only two colors are needed as all finishing ends can be one color and all starting ends of a different color. As an example plain white sleeving can be used for starting ends, with blue for the finishing leads. This sleeving is not wasted as it can be left on the wires to form an extra layer of insulation going to the commutator where the wires are bunched.

## CONTINUOUS WIRE SYSTEM

One advantage of the continuous wire system in winding loop wound armatures lies in the fact that grounds can be instantly detected if the beginning of the winding is connected to one side of a test lamp or buzzer, and the shaft is connected to the other side. As soon as a ground appears the lamp will glow, or the buzzer will indicate the trouble, and the winder can remedy the fault at once before covering the place with more turns. See Fig. 8-7.

While the armature coils of the higher voltage machines, with certain exceptions, usually contain a large number of turns of relatively small wire, this fact is not true with machines operating on 32 volts and less. Here we most often find very few turns per coil, in the neighborhood of 6 to 10 or so, and the wire possibly between number 12 to 18 gauge. An individual coil on an average armature of this type will be found to contain anywhere from 6 to possibly 10 feet of wire. Hence many rewinders resort to a scheme of winding that eliminates all bother of counting turns.

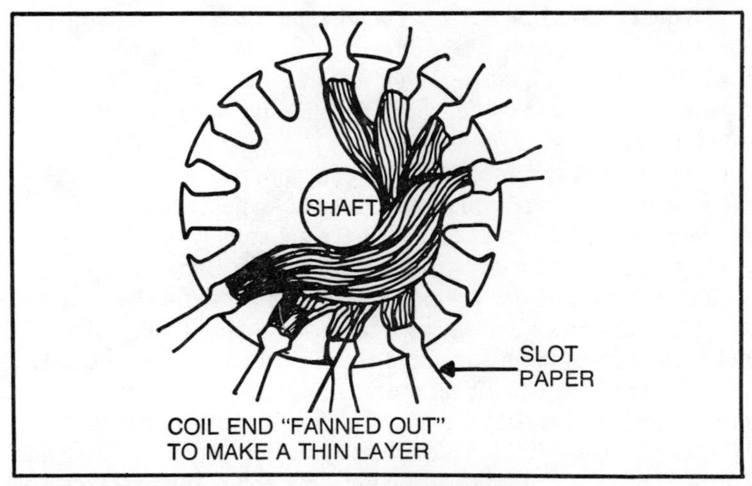


Fig. 8-6. Spreading out coil ends prevents bulkiness.

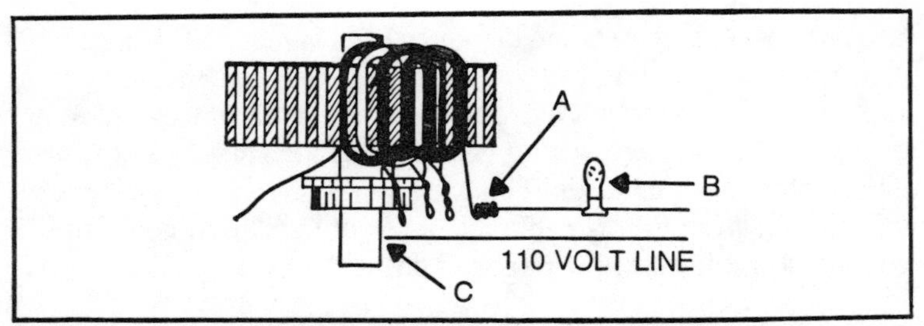


Fig. 8-7. When a ground occurs in winding, the test lamp lights at once.

The simple method by which this is done is to measure the wire from one of the original coils, and then to cut a bundle of new wires about six inches longer than the sample. This extra length, of course, is to compensate for inequalities in winding and to allow plenty of length for the commutator leads. The winder then draws wires from the bundle and winds turns into the slot until all of the length has been placed. As mentioned before in connection with this type of winding, the last few coils will need slightly more wire than will the first coils, due to the greater distance the last coils have to travel over those beneath.

This method of winding will be found very handy for the man who is constantly called away from the winding bench to answer telephone calls, or to do other work. The freedom from having to count or to remember the number of turns made will allow the winder to get back on the job with the least loss of time and the fewest errors.

When connecting any type of winding to the commutator it is found to be very handy if all of the finishing ends of the coils can be made to come out on the top layer, just under the wedge. This gives a pleasing contour to the finished job and is the way most armatures are wound.

Taking a 14 slot armature as an example, with a pitch of 1 and 7, we find that the first 6 coils wound lie wholly in the bottom of their respective slots. At this point we have six coils with 12 coil sides occupying 12 slots, with two slots still vacant. Our next coil to be wound will start in slot 7 which already holds the finishing side of the first coil wound, and we have to do something about it if we do not wish to cover up the finishing end of the coil starting in slot 1.

What is done is to lift out the finishing end (or ends if more than one in hand) and bend it back away from the rear of the armature. We now wind in the coil that takes position in slots 7 and 13. After this coil is in place we bring back the finishing end of the first coil that was in slot 7, and which was temporarily removed, and replace it in the top of slot 7 *above* the last wound coil. Thus we have the finishing lead from the lower coil coming out on top of the top coil. In the armature we are considering this must be done in each slot from No. 7 to No. 14, Slots No. 1 to No. 6 will have the finishing leads coming out on top as a matter of course.

Loop windings can be wound either right or left hand without affecting armature rotation or polarity. By right or left hand is meant the direction in which the coils are placed one upon the other. The winder can wind either way (from him or toward him) and the coils can be placed on either side of the armature shaft in making the circuit from slot to slot.

The winding pitch of an armature is found by counting the number of slots spanned by one complete coil, including the slots in which the coil rests. In the fourteen slot armature we have been considering the pitch is 1 and 7, meaning that one side of the coil is in slot No. 1 while the other side is in slot No. 7. For a 2 pole machine of this type the full pitch would be 1 and 8, or in slots exactly opposite each other. As the pitch in this case is one slot less than full pitch it is commonly known by the winding term "chorded." In a machine of four or more poles the "full pitch" would be the number of slots necessary to span the space between exact centers of adjacent poles.

# Chapter 9

# Rewinding Fan Motors

Many different kinds of electrical appliances are brought to the average electric shop for repair. A very fair proportion of such appliances needing service will be fans of one kind or another, and other small single phase motors used in the home, shop or office.

The profit from this kind of work depends almost entirely upon the judgment and experience of the service man. Judgment must be used to decide whether it will be best to take the job in for repairs, or whether the customer should be sold on the purchase of a new and better fan, vacuum cleaner or other appliance. Each case must be considered on its merits, but as a general rule it is poor policy to accept a repair job if the estimated cost of reconditioning a small appliance exceeds 50% of the price of a new article. Unless, of course, the customer insists that the work be done regardless. Such a policy leads to more sales, better profits, and in the long run to more satisfied customers.

A certain portion of the market has been flooded with cheap merchandise. Obviously it is impractical to repair and impossible to make any profit from this kind of business, and the sooner such items land in the junk box the better for all concerned. Experience will teach a service man that many inexpensive appliances should be treated with a hands off policy when it comes to the shop for repairs. This is because parts are hard or even impossible to obtain, and the amount of labor involved would be out of proportion to the value of the unit.

Many electrical repairmen, while familiar with the operation of small single phase motors above the 1/8 horsepower size, hesitate to dig too deeply into the miniature single phase motors because they are not familiar with the various starting and speed control circuits used on so many fans and other appliances. Some of these circuits seem quite complicated when seen in the final assembly, and are often very hard to trace out if no diagram is at hand. As most repair men know, it often takes ten times as long to locate the trouble as it does to fix it when once found.

The purpose of this chapter is to explain, and show by simple diagrams, the circuits commonly used in single-phase fan motors, especially in the older models. Because of space limitations the exact circuits for all makes and models cannot be given, but it is believed enough are described here to give a working knowledge of the principles involved.

## EARLY FAN MOTORS

The earliest type of alternating current fan motor was copied from the direct-current fans in use at that time. It was discovered that a series type motor would run on ac as well as dc, with the exception that the use of ac current caused excessive sparking at the brushes. This complaint was diminished somewhat by making slight changes in the winding of the armature and field windings. There remained, however, the matter of speed control. On the earlier motors this was accomplished by shifting the brushes from the neutral position, as indicated in Fig. 9-1.

In later types, many of which are in use today, the speed of the fan is controlled by inserting a resistance in series with the motor circuit, thus lowering the voltage applied to the motor windings.

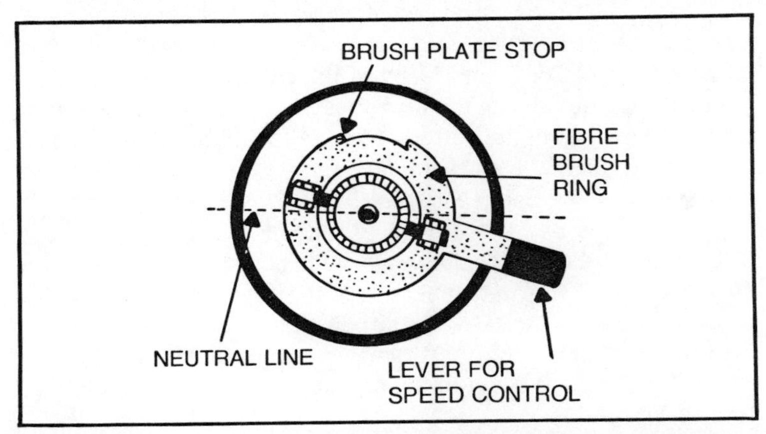


Fig. 9-1. Old style fan motor of the series type with speed regulation obtained by shifting brushes.

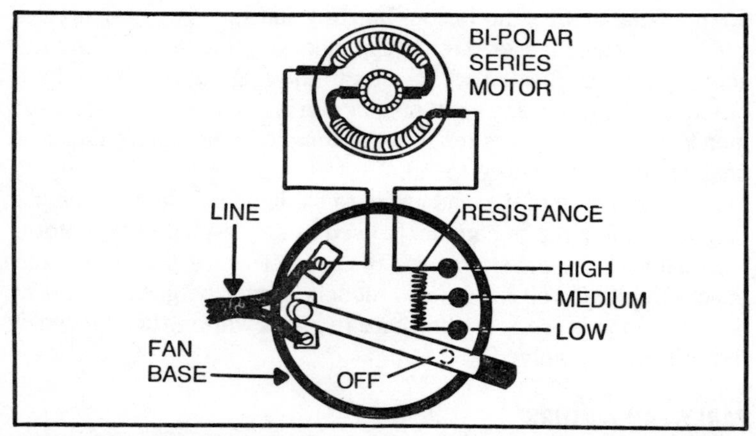


Fig. 9-2. Circuit of a series-commutator fan motor with speed control by a tapped resistance.

This, of course, reduced the speed of the armature in proportion to the value of the resistance. Many of these fans have a single resistance, giving a high and a low speed; others use a tapped resistance giving three or four speeds. A diagram of the circuits of a simple three speed fan of the commutator-series type is shown in Fig. 9-2.

## INDUCTION MOTORS

The next step in the advancement of fan motor design was the use of an induction motor of the simplest type. These motors made to operate on 60 cycle current employ four poles to obtain the most satisfactory speed for the most common type of fan blades—in the neighborhood of 1600 rpm. Many of the cheaper fans used a cast field ring, but because of the heavy losses common to this type of construction most of the better fans used a laminated field frame.

Since induction motors are not inherently self starting, some means of causing a displacement of the pole flux had to be provided to cause the rotor to start revolving. One means of accomplishing this was by shading the trailing edge of the poles, thus retarding the flow of magnetic flux in that section of the pole surrounded by the copper band, while in the rest of the pole the flux is in phase with the field current. Shaded coil construction reduced manufacturing costs and will be found in many of the cheaper induction motors in this type of service.

Induction type fan motors use a squirrel-cage rotor somewhat similar to the rotors used in larger induction motors. The copper squirrel-cage construction of the typical fan rotor, however, has a higher resistance than that of those used in motors for power drives. This higher resistance of the squirrel-cage rotor is necessary where the field strength is to be reduced in the controlling of fan speed. As the voltage applied to the stator winding is reduced it decreases the density of the magnetic flux and allows greater slippage of the rotor.

The speed of shaded coil motors is usually controlled with the aid of a choke coil or resistance placed in series with the main field. A diagram of a typical single-phase fan motor of the shaded pole type, one employing a resistance unit for speed changes, is shown in Fig. 9-3.

A common type of ceiling fan is similar in principle to the type just described. Some ceiling fans make use of copper band shading on the individual poles while others use a continuous shading coil that is wound through the slots on top and isolated from the main winding. The terminals of this shading coil are soldered together to make a complete circuit having no connection with line voltage at any point. The continuous wire would shading coil balances the retarded flux in all poles alike, and makes for a smoother and more quiet running motor.

Because of the lower speed at which they operate, all ceiling fan motors have a larger number of poles than does the conventional desk type of fan. When ceiling fans become noisy in operation the

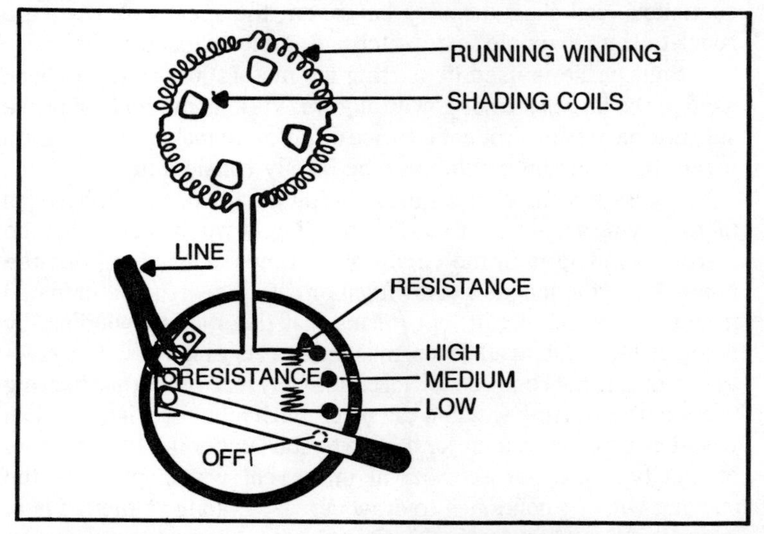


Fig. 9-3. Single-phase motor of the shaded coil type. A tapped resistance controls speed.

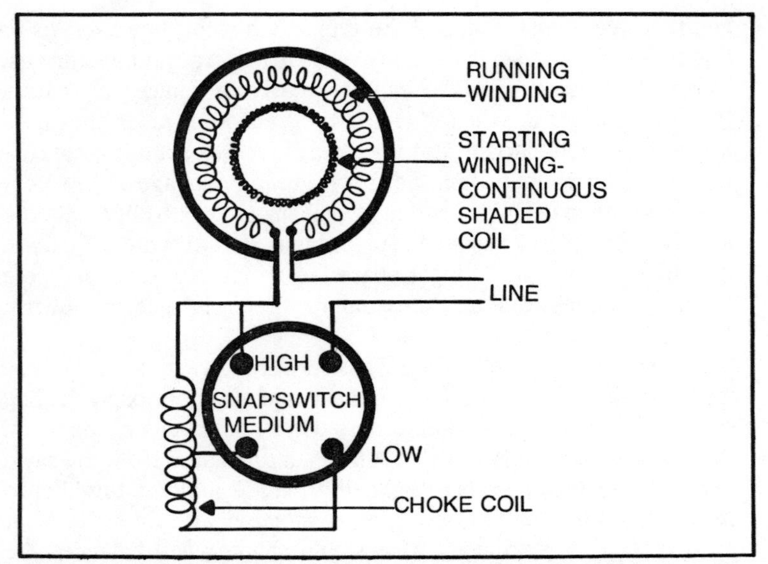


Fig. 9-4. Circuits of a ceiling fan motor using a wire wound shaded starting coil with choke coil for speed control.

trouble can usually be traced to lack of lubrication or wear in the rotor bearing. If excessive, the latter condition may allow the rotor to scrape against the poles. Figure 9-4 shows the circuits of a continuous coil shaded ceiling fan motor, the speeds of which are regulated by a four pole snap switch and a center tapped choke coil.

Split-phase is used in starting fan motors just as it has been used in the larger size of power motors. Various methods of phase splitting have been brought into use on different makes of fans, and a few of the common circuits will be briefly considered.

Figure 9-5 shows the simple circuit details of a splitphase fan motor having a main and a starting winding. It will be noted that the starting winding is in the circuit at all times. This simplifies the operation of the fan, as a centrifugal or cutout switch is eliminated from the starting circuit, but means that the starting winding, by being in the circuit at all times, must have a very high resistance. As a fan motor is not required to start under any load other than bearing friction, the starting winding can be of much higher resistance than would be possible in a motor having to start under full load. Motors of this type are not economical in current consumption as the starting winding continues to draw current as long as the fan is in operation. Speed control in this type of fan is usually by a resistance coil having one or two taps.

## SPLIT-PHASE MOTORS

In Fig. 9-6 we have a split-phase fan motor circuit very similar to the one in Fig. 9-5, except that a centrifugal switch has been placed in series with the starting winding. Here the starting winding would be heavier wound as it is in use only during the starting period. Having less resistance it will aid the rotor to attain running speed in a shorter period of time and, while it will consume more current, this can be disregarded due to the short space of time the starting winding is brought into use. A choke coil in the fan base controls the voltage impressed on the windings and gives a number of speeds corresponding to the number of taps provided.

A very efficient type of desk fan that has been produced in large numbers is shown in Fig. 9-7. This type of fan circuit is familiar to most service men as it is one of the types having a three conductor cable connecting the motor with the base. The motor winding circuit is like that of Fig. 9-6 with the exception that the centrifugal cutout switch is not used, one end of the starting winding being directly connected to one end of the main or running winding. The features of this design are a high starting torque, better than average efficiency at running speeds, and good speed control.

A transformer is to be found in the base of a motor of this type. The transformer primary is tapped at several points and is placed in series with one lead of the main, or running winding. The trans-

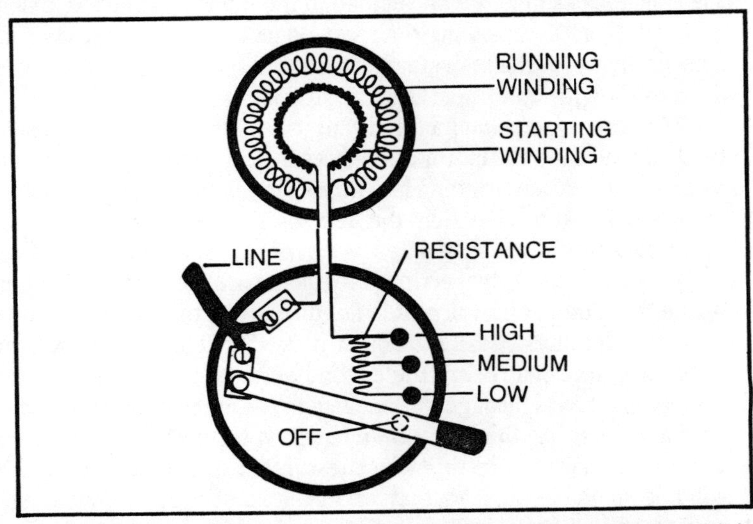


Fig. 9-5. Split-phase fan motor with high resistance starting winding, which is permanently connected in the circuit.

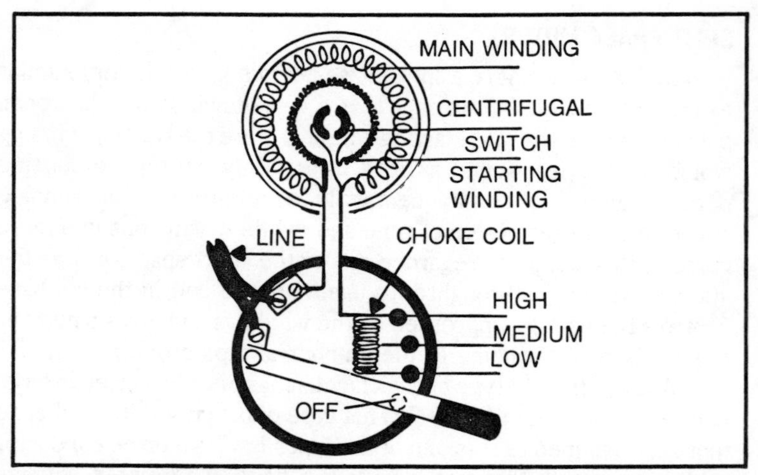


Fig. 9-6. Circuits of a split-phase fan motor using a cutout switch in series with the starting winding.

former secondary is connected to the motor side of the primary at one end, and to the free end of the starting winding on the other side. When a circuit of this type is thrown on the line the current for the main winding has to pass through the transformer primary, and the voltage induced in the secondary winding of the transformer is fed to the starting winding circuit. The voltage lag in the secondary of the transformer plus the normal lag in the high resistance starting winding gives a phase-angle almost equal to a two-phase motor. This accounts for the high starting torque in this type of fan, and has a lot to do with quiet operation at all speeds.

Another type of fan motor circuit using a three conductor cable between the base and motor eliminates the use of a starting winding by dividing the main winding into three parts, and placed in relationship to each other so that the flux of the field has a rotating characteristic closely associated to that of a three phase motor. One end of each of the three sections of winding are connected together inside the motor. Line current is fed directly to the free end of one winding, and after passing through it, is distributed into the two remaining sections. From the second section of the winding the current must pass through a resistance to reach the other side of the line, and from the third section must pass through a tapped reactance or choke coil. In this way the current in one section of the winding is made to lead that of the next section (because the resistance and choke coils are of different values) and the rotor is caused to revolve in the field.

Speed control in a motor of this type is obtained by adding more or less turns to the working part of the circuit that includes the resistance and choke coil. When the switch is moved to the medium or low positions more turns are added to the choke coil and to the resistance, with the result that the voltage applied to the motor windings is reduced, and consequently the speed of the rotor. Any reduction of voltage in the two sections of the winding connected to the resistance and choke coils is automatically reflected in the section of the winding connected to the line, as this winding is a common path for the other two sections. A schematic diagram of a fan having a circuit of this type is given in Fig. 9-8.

## SERVICING PROBLEMS

In the servicing of electric fans the most common complaints are worn out bearings, stripped gears on oscillating types, and burned out resistance or choke coils. As a general rule the motor windings give little trouble except where the impressed voltage has been too high, where the fan has been connected to current of the wrong number of cycles, or where the size or number of the fan blades has been changed. Fan motors are designed to operate at

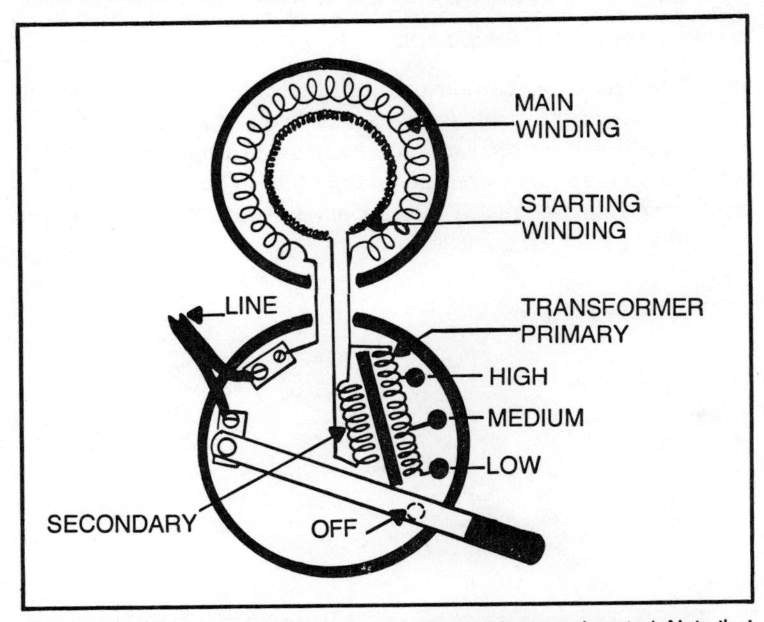


Fig. 9-7. Split-phase fan motor using a transformer for speed control. Note that the starting winding is connected to the transformer secondary.

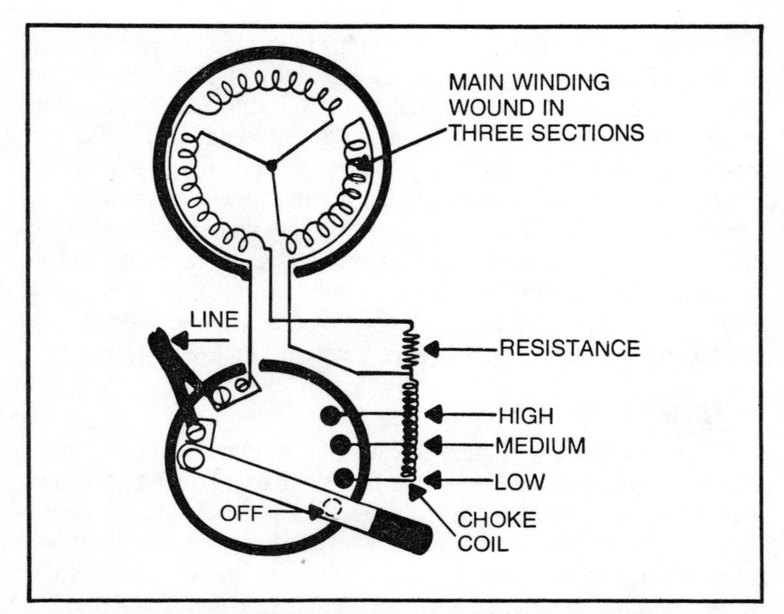


Fig. 9-8. A split-phase fan motor with winding in three parts. Each section of the winding is spaced one third of a pole apart around the stator instead of as shown in this schematic diagram. A motor of this type has many of the operating characteristics of a three-phase motor.

certain speeds under certain load conditions. When larger blades, or more blades of the same size, are attached to the rotor shaft, the rotor speed will be reduced on the one hand, while current draw from the line will be increased on the other. The same conditions apply when fan motors are used for other purposes where the load or speed is changed greatly.

# Chapter 10

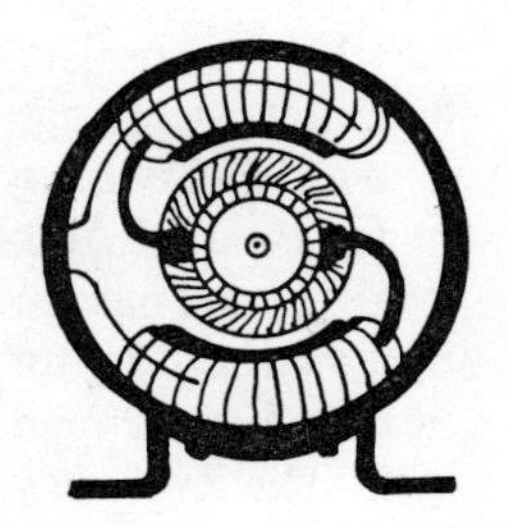

# Rewinding Automotive Armatures

Many requests have come from motor shops asking for general information on the winding of low voltage armatures such as are used in automobile and marine engine generators and starters. Because it would be impossible to answer all these queries in detail, we are giving here a number of suggestions and instructions for rewinding six and twelve volt armatures.

Before going into the methods used in this branch of rewinding, let us first consider the possibilities for making profits from this sort of work. In the larger cities the automotive field is keenly competitive and in many cases prices have been hammered down to bed rock. For this reason alone the well established motor shop will not find this form of rewinding attractive, and the best interests of all concerned will probably be better served if the motor shop sticks strictly to its own field.

In the larger cities there will usually be found one or more attractive shops that are on a quantity production basis, and who, by the use of cheap, unskilled labor, are able to turn out the work at extremely low prices. Not only do they use cheap labor but in most cases they use undersize wire and many other short cuts to beat down the price. No thought is given to quality or lasting service.

At current prices the most popular types of automobile generator armatures, such as Ford and Chevrolet, can be bought at low prices. Obviously, considering even the cheapest labor and materials, there can not be much profit. The question is often asked: How do they do it? A visit to one of these production shops will supply the answer.

I recently inspected such a shop. Eleven boys and girls and one experienced armature winder were employed in the work of turning out rewound armatures. One young boy, fresh from the farm, did nothing all day long but strip old cores. After stripping, the cores were placed in a gas oven to burn out the remaining paper and dope. Another boy replaced defective commutators or cleaned up the old ones for rewinding. One girl cut and inserted the slot insulation. One girl and two boys wound the coils in the slots, leaving the free ends extended. Another boy placed coil leads in the commutator bar grooves, and passed them on to another boy who did the soldering. Still another boy turned the commutators on the lathe and did the final testing. Dipping, baking and undercutting were done by two girls. The foreman was kept busy supervising the work and attending to the few jobs that needed special knowledge.

In this shop not one employee, except the foreman, was capable of winding a complete armature. Each one had been trained to do but one thing, and handled the same operation day after day. The capacity of this shop was about 300 armatures per day, and it is easily seen that the skilled armature winder can not compete in this market.

So far, from the standpoint of the skilled workman, we have painted the picture rather dark. There is, however, another side of the picture if we move to one of the smaller towns some distance from a large city. Here, through automotive jobbers and auto supply stores we still have the same competition in the small car generator armature field, but these stores are not likely to carry armatures to fit old, obsolete or high priced cars, and these are the type that net the rewinder real money.

Automotive armature winding fits in well with the small town motor shop because the same equipment and stock of materials can be used for both types of work. Low voltage armatures are easier to wind, especially for the beginner, because there are fewer turns, fewer coils, and less attention need to be given to insulation. The wire sizes used in automotive generator armatures are the easiest for hand winding, as they are neither too small nor too stiff. Because of the large demand, commutators, fibre end laminations, etc., are quite low in price.

If the rewinding of automotive generator armatures is to be a spare time business for the motor shop, a representative stock of rewound cores must be kept on hand so that exchanges can be made without delay. The exchange method allows prompt service to the buyer (who is invariably in a hurry) and gives the shop the opportunity of doing the winding when other work is not pressing. It is understood that odd types and obsolete armatures can not be stocked, but as these command much higher prices the shop can afford to turn them out as special jobs.

The best way for a shop to build up a stock of automotive armature cores for rewinding is to patronize a local junk yard. Cores in good condition, but with burned out windings, can usually be obtained easily. In selecting cores, particular attention should be given to the condition of the laminations, trueness, threads, centers and to the bearing surfaces. It will not pay to rewind cores that are not in good condition.

## WINDING DATA

Just as in the winding of any other type armature, the first step is to secure the winding data from the original winding, and to set it down on a form card which can be filed for ready reference. In this manner a more or less complete record of all types of armatures can be built up, after which most of the time that would otherwise be spent in tracing old windings can be saved.

If the shop does a sufficient volume of armature winding it will pay to use a burning oven to aid in the stripping and cleaning of old cores. After a brief period in an oven of this type all organic matter is completely burned and the wire and insulating material are easily removed. If the commutator is to be used again it should be removed before the burning process. Many of the commutators used in late model armatures use bakelite insulation between the bars, and this is easily ruined. In fact, the great majority of bakelite insulated commutators are badly charred when the armature comes in for rewinding, due to overheating in the generator, and will have to be replaced.

## REMOVING COMMUTATORS

Before removing commutators, the distance they are set back from the bearing shoulder of the shaft should be measured, and the new one pressed on the same distance. Before installing a new commutator, or an old one, the fibre end or star washer should be fitted on the shaft, and the slots in the commutator bars should be scoured bright to make soldering easy.

A special grade of slot insulating paper suitable for almost all automotive armatures can be purchased in 10 pound rolls. This paper is about the right width for most generator armatures, and needs a minimum of trimming. A roll of crepe paper about 1 inch wide is handy for banding exposed parts of the armature shaft, and for insulating the armature neck between winding and commutator leads. Crepe paper can easily be formed to fit over irregular and tapered parts of the winding.

With the exception of the Ford model T generator, most automotive armatures are wound with two wires in hand. It simplifies matters a great deal and tends to prevent mistakes if one of the wires carries a tracer. In making commutator connections the winder can then place the ends alternately white and tracer, white and tracer completely around the commutator. Number 17 sce is the size most used in automotive work.

In order to get a symmetrical winding it is best to place all starting leads into the right commutator bars as the coils are wound, leaving the finishing ends free to come out on top of the finished winding. Care must be used to get the right commutator pitch or the armature current will be reversed. Commutator pitch is the number of bars that the winding advances. A change in pitch of one bar will cause the winding to be progressive or retrogressive as the case may be. To explain this matter more fully we will consider a wave winding (used in a four pole generator) having 21 slots and 21 bars.

In Fig. 10-1 we find the pitch is from bar No. 14 to bar No. 3 in tracing to the right, which gives us an advance of 10 bars, as we do not count the bar in which the coil started. As we wind a second coil, also advancing 10 bars, it brings us around to bar No. 13, which is just behind No. 14, where we started to wind. Then, because the winding has dropped back one bar in tracing to the right, we have a retrogressive winding.

If, however, we made our connections to the commutator as indicated by the dotted line, advancing 11 bars instead of 10 each time, we would discover that instead of coming around to bar No. 13, the winding would cross over the lead to bar No. 14 and enter bar No. 15 making a gain of one bar in tracing through the two coils that make one path around the armature. Because we have gained one bar we call this a progressive winding.

If an armature is wound progressive when it should be retrogressive, or retrogressive when it should be progressive, it will motor in the wrong direction when tested inside the generator, and it will not charge if driven in the original direction. The armature can be made to charge in the original direction of rotation if the field leads are reversed, but such a change is not always easily made in an automobile generator.

Most of the modern generators, being 2 pole machines, are lap wound, and two forms of winding are in current use, hand wound and form wound. In the former, also called a "balanced" winding, the coils are wound on a form and then assembled in the armature slots. This type of winding has a better mechanical and electrical balance because all coils have the same length and weight of copper, and hence, the same resistance. Two disadvantages of this kind of winding, from the viewpoint of the small winding shop, are that it is much harder to wind in small quantities, and that in working the coils into the slots there is always considerable danger of breaking through the insulation. Since straight hand winding can be substituted satisfactorily for the original form winding, most shops use the latter method in rewinding cores of this type.

The key to a good hand winding lies in the formation of the first three coils. These first three coils determine the shape of the completed winding as all subsequent coils wound must pass over them. The first coils wound should be made to hug the curve of the armature shaft and the core ends. It is necessary to tamp the wires down into the bottom of the slots and keep them under reasonable tension. Coil ends can be tamped into shape by means of a smooth wooden block and a light hammer.

In hand winding, the first few coils wound will occupy the bottom position in the slots on both legs, while the last few coils will occupy only top positions. In the latter case, the finishing ends of the coils will automatically come out on top of the winding, where we want them, but some special provision must be made to bring the finishing ends of the lower coils to the top. One method of doing this is to remove the last half turn of the finishing ends from the slot before a top coil is wound over it. Then at the completion of the top coil, and before the wedge is driven, the finishing ends of the lower coil can be brought upward and placed on top. Since there will be one set of finishing ends per slot regardless of the position on the armature, there will be no chance of a mistake with this method.

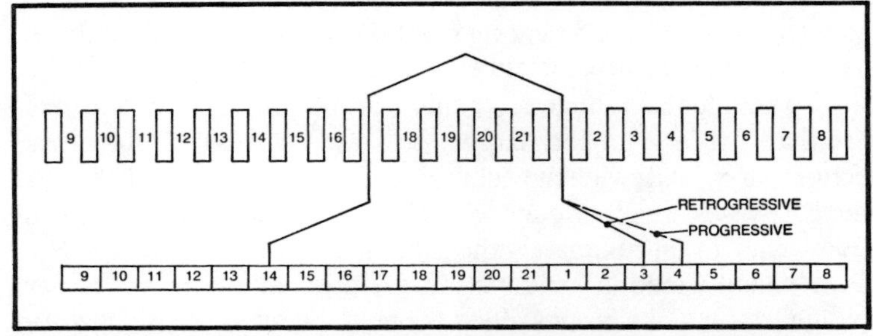


Fig. 10-1. An example of an accurate diagram of a wave winding. The connections for both a retrogressive and a progressive winding are indicated by the solid and dotted lines.

## OLDER GENERATORS

There are a few armatures in the automotive field that do not follow conventional practice. Some of the older Hudson-Essex generator armatures were wound with one dead ended coil to mechanically balance the winding. In other words, when winding the armature with two wires in hand there would be one individual coil left over for which there were no bars on the commutator. The ends of this coil must be cut off and the coil left open circuited. Generator armatures on the older Packards use 20 slots and 41 commutator bars. Since they were wound two in hand there was one odd commutator bar left over. To complete the electrical path of the winding it was necessary to cross connect one pair of bars on opposite sides of the commutator. This and other typical types of automotive windings are shown in some of the accompanying charts.

A very large number of the cheap replacement automobile generator armatures on the market are wound with wire that is one size smaller than originally specified. Rewinders sometimes fall for the temptation of cheaper material and labor costs, but the shop that intends to establish a reputation for a good product should avoid this pitfall. The use of smaller wire gives a higher internal resistance to the winding and will cause the generator to heat up more rapidly. If fewer than the usual number of turns are used, even of the proper size wire or larger—the result will be to lower the efficiency of the machine, and a higher speed will be required to obtain a normal charging rate.

The rewinding shop is occasionally called upon to rebuild or convert an old generator to a different purpose. Wind chargers, arc welders and 110 volt generators for public address systems are some of the common uses of old automobile generators. The change from 6 to 12 volts, or from 12 to 6 volts is one that is often requested. Some experimenting must go along with many of these changes, but there are certain rules that can be followed.

To double the voltage of a generator the rule is to use a size of wire that is three numbers larger. This wire will be half the cross sectional area, and twice the number of turns will be used. Thus, if a field coil was originally wound with 100 turns of No. 18 wire, it could be rewound, in approximately the same space, with 200 turns of No. 21 wire, and would be suitable for operation at twice the former voltage. It is not always possible to double the turns on an armature for various reasons, but if the original winding is replaced with one of one or two numbers larger, and all available space is filled, the results will usually be satisfactory.

Other operations on the rewinding of automotive type armatures, such as soldering, turning, undercutting, dipping and baking follow standard practice and offer few obstacles. Many motor shops have taken up automotive rewinding as a side line, and have reached the point where this class of work carries much of the overhead through dull seasons. Since it is a branch of the business that demands little additional investment, and since there is a fairly constant demand, it is something to consider.

# Chapter 11

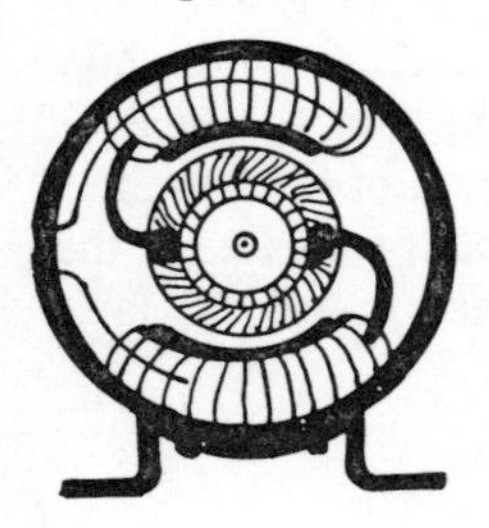

# Rewinding Small Polyphase Motors

The single phase motor, because of its wide use in domestic and commercial applications, is, and perhaps always will remain, the principle source of revenue for the small motor repair shop. When looking inside a single phase motor, it is extremely easy to identify the coils and poles, and their polarity with respect to each other. The speed of the motor can be told as well, just by looking at the windings. One can understand quite easily the methods for rewinding such a motor, since it is easy to take apart, its internal structure being so orderly.

Looking inside a polyphase motor, however, is an entirely different matter. One does not see a definite grouping of coils, but instead, a continuous pattern of shapes and forms which hide all the information visible inside the single phase winding. In fact, a polyphase winding may not even appear to have any poles at all, just a continuous weaving of the wire. After some examination, it may become apparent how each coil is shaped, but there are still many mysteries which go unsolved.

Many winders who have mastered the technique of the single phase motor hesitate to branch out into the two and three phase field for the simple reason that so much of the latter work is large and heavy. While single phase motors are usually smaller than 10 horsepower, polyphase motors come in all sizes. Also, due to the reasons mentioned above, rewinding these motors requires some specific knowledge of the types of windings.

Small motor repair shops should give some serious consideration to the idea of rewinding polyphase motors, as there will always be a portion of the customers who seek this work. If a large motor is too big to handle in the small shop, the serviceman can refer the customer to a larger shop. In the home, rewinding polyphase motors can be quite rewarding and educational, broadening the mechanic's understanding of motor operation.

Numerous requests have been made for some information regarding the rewinding of the smaller two and three phase motors. Most of these requests have come from shops in smaller cities and towns where an additional job or two often spells the difference between a good and a bad week. Customers living in a larger city usually head straight for the big shops, but may be turned down if the motor is too small for them to handle. In most localities, the percentage of two or three phase motors is small and the fractional horsepower group is small. In fact, two phase motors are so scarce that it can safely be said that the small shop would see only three phase motors in the smaller sizes.

Most of the people who would like to get into the rewinding of these polyphase motors are usually quite experienced at rewinding single phase motors, but want a few more pointers before venturing out into three phase work. Describing the types of windings used is the best way to teach someone the nature of the work, and the theory behind it. Before each type of winding is examined, the general statement could be made that: A three phase motor is essentially three separate, overlapping, single phase windings.

One way to show a three phase winding is shown in Fig. 11-1. Notice that this is really three single phase windings connected in a star or Y pattern. In this diagram, there is no overlapping of coils. While it is entirely possible to produce a winding like this, overlapping the coils would allow for greater size of each pole and a more efficient motor. The way the coils are shaped and overlapped is what gives each winding its name. The following types of windings are used on nearly all polyphase motors:

- Basket winding (one layer)
- Pulled diamond coils (two layer)

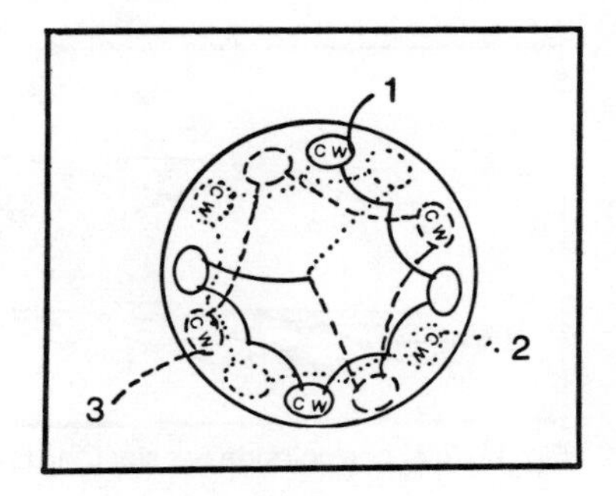


Fig. 11-1. A three phase, 4 pole winding. Poles marked "cw" are wound clockwise, while unmarked poles are wound counterclockwise.

Fig. 11-2. Left, a one layer winding. Right, a two layer winding.

- Flat diamond coils (two layer)
- Copper strap winding (usually two layer)
- Chain winding (one layer, rarely used except for large, slow speed motors)

Before any mention on exact winding types, it must be understood what is meant by "one layer" and "two layer" windings. All this refers to is how many coil sides there are in each slot (see Fig. 11-2). In the one layer winding, there is only one coil side filling the whole slot. In the two layer winding, the bottom of the slot is filled with the side of one coil, and then another coil side is placed on top of it. Figure 11-3 shows the general lacing of both one and two layer windings. Note that the one layer winding only has one coil in each slot, while the two layer has two. While single phase motors are made of concentric coils in the windings, three phase motors are made up of overlapping coils, as shown in Fig. 11-4. There may be any number of coils in each pole.

The basket winding is so named because the appearance of the coils outside the slots resembles that of the weaving of a basket. It is extremely popular in both small and medium sized motors. Basket wound coils are all uniform in size and shape and are very easy to produce. The following conditions are required for a basket winding:

- The total number of slots must be an even number.
- The number of slots spanned by each coil must be an odd number.

Again, the basket winding is a one layer winding and the total number of coils is half the total number of slots.

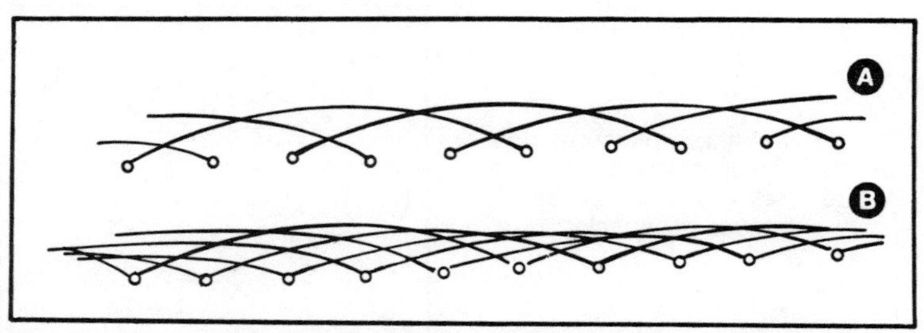


Fig. 11-3. A, one coil side per slot (one layer) B, two coil sides per slot (two layer).

## COIL SPAN

On the subject of coil span, examples would be 1 and 6, 1 and 8, 1 and 10, etc. To get an odd span the last number must be even. For example, let us consider a 48 slot stator that is to be wound for 4 poles and three phase operation. To obtain a good coil span, one can divide the number of slots by the number of poles, providing that the result is a whole number. If it is not, an approximation will do. In the above example, 48/4 = 12. We will choose a coil span of 1 and 12 and check it out on a penciled chart to make certain it will work. A simple chart consists of the numbers from 1 upwards written on a line across a sheet of paper, such as shown in Table 11-1. A study of the chart will show that one slot is skipped in the beginning of each coil so there will be a vacant slot available for finishing the side of some other coil. The chart proves that the winding will work out correctly in as much as we find that whenever we have either a top or a bottom coil side marked above or below a slot number, there is none charted on the opposite side of the slot number. This shows that there is only one coil side per slot. Thus slot #12 has the top side of coil 1, slot #13 has the bottom side of coil 7, and so on.

Table 11-1(B) shows what would happen if the pitch was 1 and 9, forming a coil span of 8, which is an *even* number. Looking at the chart, one can see that slots 9, 11, 13, etc., contain two coils while the even numbered slots would contain none. This would be a form of a two layer winding, which is not what we're after.

The wooden practice block described in Chapter 3 is another way to check and see if a coil span will work. It also gives a visual indication of how the finished winding might look.

To make the coils for a basket winding, wire may be wound around pegs spaced apart on a board, or a winding machine with a

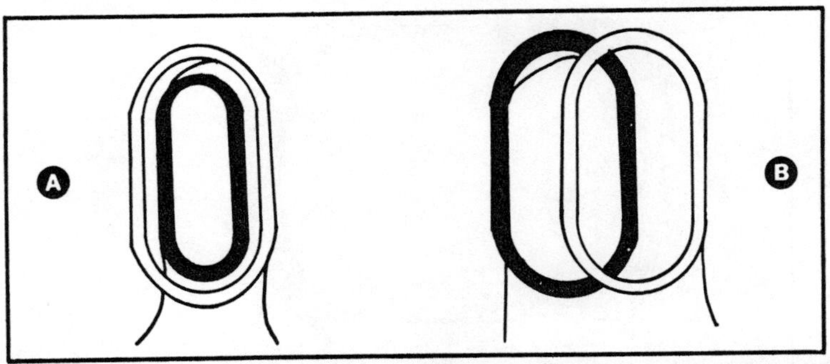


Fig. 11-4. While single phase motors use concentric coils for their windings (A), polyphase motors use overlapping coils, (B).

**Table 11-1. Charts to Plan Out A Winding, Shows a Coil Span of 1 and 12, While B, Shows a Span of 1 and 9.**

| | | | | | | | | | | | | | | | | | | | | | | | |
|---|---|---|---|---|---|---|---|---|---|---|---|---|---|---|---|---|---|---|---|---|---|---|---|
| Top of Coil No. | | | | | | | | | | | | 1 | | 2 | | 3 | | 4 | | 5 | | 6 | |
| Slot No. | 1 | 2 | 3 | 4 | 5 | 6 | 7 | 8 | 9 | 10 | 11 | 12 | 13 | 14 | 15 | 16 | 17 | 18 | 19 | 20 | 21 | 22 etc. | A |
| Bottom Coil No. | 1 | | 2 | | 3 | | 4 | | 5 | | 6 | | 7 | | 8 | | 9 | | 10 | | 11 | | |
| Top of Coil No. | | | | | | | | | 1 | | 2 | | 3 | | 4 | | 5 | | 6 | | 7 | | |
| Slot No. | 1 | 2 | 3 | 4 | 5 | 6 | 7 | 8 | 9 | 10 | 11 | 12 | 13 | 14 | 15 | 16 | 17 | 18 | 19 | 20 | 21 | 22 etc. | B |
| Bottom Coil No. | 1 | | 2 | | 3 | | 4 | | 5 | | 6 | | 7 | | 8 | | 9 | | 10 | | 11 | | |

Charts such as these are useful in determining the proper coil span for a given stator when a basket type winding is to be used. The upper chart indicates that a coil span of 1 and 12 will be practicable for a 48 slot stator to be wound for four poles, three-phase. The lower chart indicates that a coil span of 1 and 9 will not be permissible. . . . . . . . . . . . . . . . . . . . . . . . . . . . . . . . . . . . . . . . . . . . . . . . . . . . . . . . . . .

suitable mold may be used. The correct number of turns are wound, the coil is tied to keep it together, and it is removed from the form. A typical basket winding coil is shown in Fig. 11-5. The slots are laid with an insulating strip of cloth, paper, or plastic, and the coils are inserted. Any slot may be called #1, with the others counting away in either direction. The first slot in which the coil is laid is the slot number shown on the chart. The other end of the coil is called the "throw" side and is swung down into the appropriate slot for the span being used. The best way to wind is *downhill*. This refers to starting the first coil as being the top in slot #1, and in the example given earlier, the bottom of the same coil rests in slot #12. The second coil is laid in slot #3 and thrown into slot #14. It will become apparent that as you progress around the stator, some of the first coils which were laid in place will have to be lifted back out of the slots to allow the final coils to be put underneath them. The alternative is to lay the first several coils down on one side only, and then close the winding when all the coils are put in place. The coils are taped outside of the slots and tapped into final shape and position

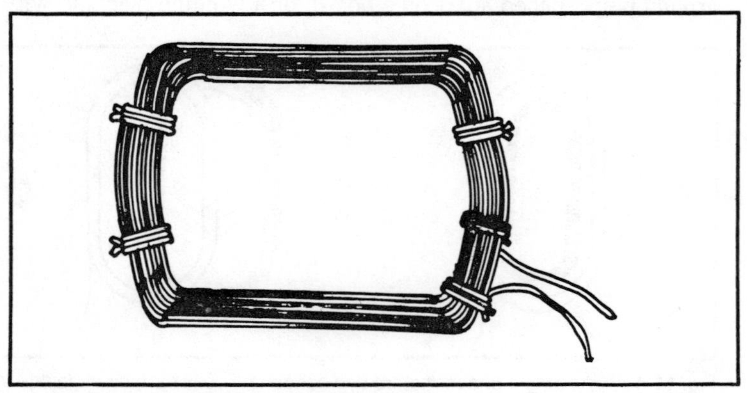

Fig. 11-5. A coil wound for a basket winding.

Fig. 11-6. Basket windings are laid in this fashion.

with a fiber drift and a rubber mallet. The coils must be all below the core so they do not rub on the rotor. The finished form of basket wound coils is shown in Fig. 11-6.

What about connecting the coil ends together? In the example winding, there are 24 coils, divided by 4 poles, divided by three phases. $24/4/3 = 2$, so there are 2 coils per pole in this winding. Using dots to represent the coils, this winding is shown in Fig. 11-7. The individual phases are then connected as single phases. Of course, the number of coils per pole does not always have to be a uniform number, such as 2. The four poles, with two coils each, in this winding might have been 2-3-2-3 in some other motor with a different number of slots or span. There will always be a nearly equal number of coils per phase, however.

Two layer windings use either flat or pulled diamond shaped coils. The pulled coil is shown in Fig. 11-8. It has two different

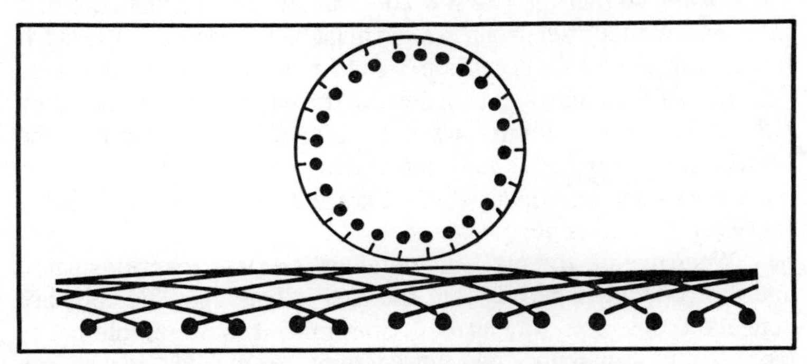

Fig. 11-7. This is how the basket windings are connected.

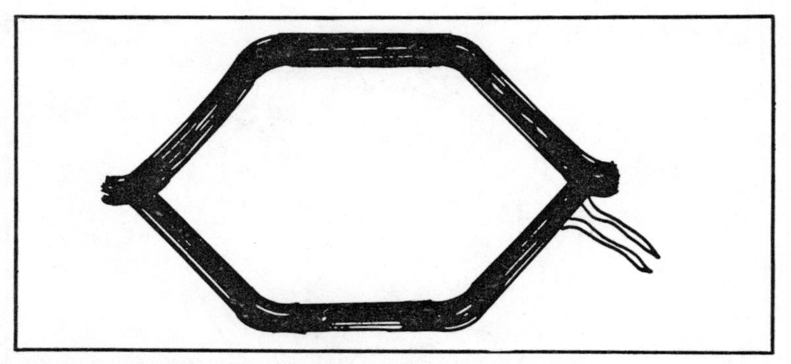

Fig. 11-8. Typical diamond shaped coil with pulled ends.

heights, which allow the coils to be spaced slightly farther apart outside the slots, permitting better ventilation. The flat diamond is similar in appearance to the pulled coil, except that it is not twisted at the ends.

To make the pulled diamond coils, they can be wound on a board of pegs like the basket winding, or, they can be wound on a machine. To twist the ends, they are clamped in a puller which bends the wires to the shape shown. While it is possible to do this by hand, especially if the wire is small, a mechanical aid is advantageous in that it allows all the coils to be uniform.

The chart which was used for the basket winding can be used to size up coils for a two layer winding, except that the coils are laid in every slot and no slots are skipped when laying the coils, except for the beginning. Figure 11-9 shows the arrangement and grouping of a two layer diamond coil winding with two coils per pole.

Note that within the two coils of any pole there are two slots. Like the basket winding, some slots must be left open when starting the winding so that the last few coils can be placed under them. If there are two coils per group, 4 slots must be left empty on the start. If there are three coils per group, 6 slots must be empty on the start. This can be accomplished by inserting the beginning coils and then pulling one side of them back out, or else by just inserting the bottom sides of the first coils and closing them only when the last coils have been laid under them. Figure 11-10 shows the start if there are two coils per pole.

When making the two layer winding, the slots are preinsulated and the coils are taped as with the basket winding. The coils are connected the same way as well. Somtimes it is preferable not to tape the pulled diamond coils, since if they are small in size it is very

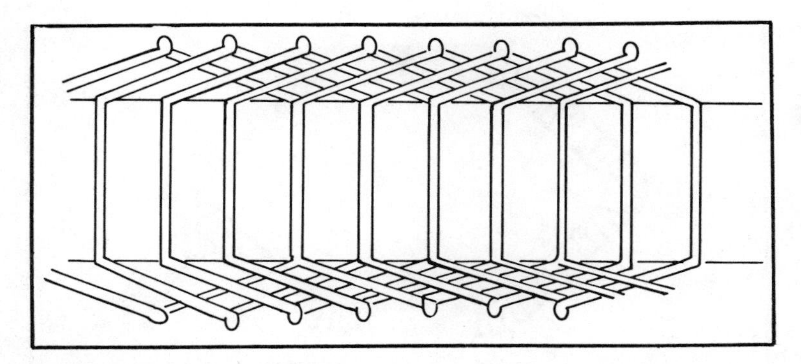

Fig. 11-9. A two coil per pole winding.

difficult to get the tape around the ends. As an added precaution, plastic or mica triangles are sometimes placed between the poles of each phase, providing that they were present in the original winding (Fig. 11-14 shows this technique used with flat diamond coils). With the two layer winding such as this, there is the same number of coils as there are slots.

Many smaller motors with two layer windings use flat rather than pulled diamond coils. These coils are preformed and partially taped before they are inserted. Figure 11-11 shows one of these coils ready for insertion into the stator. Note that the tape extends an inch or so on what will be the bottom side of the coil, while on what is to be the top side the tape lacks an inch or so of reaching the bend. The leads are brought out near the diamond nose. The reason for this will take a little explaining. Induction motors are made with either open or semi-enclosed slots. The semi-enclosed slots have very narrow openings at the tops which can only have one or two wires inserted into them at a time. When the bottom side of the flat coil is ready to be inserted, it can be turned edgewise so the narrow cross section of the coil goes in easily. When the coil is rotated down, to insert the top side, the wires must be squeezed in a few at a

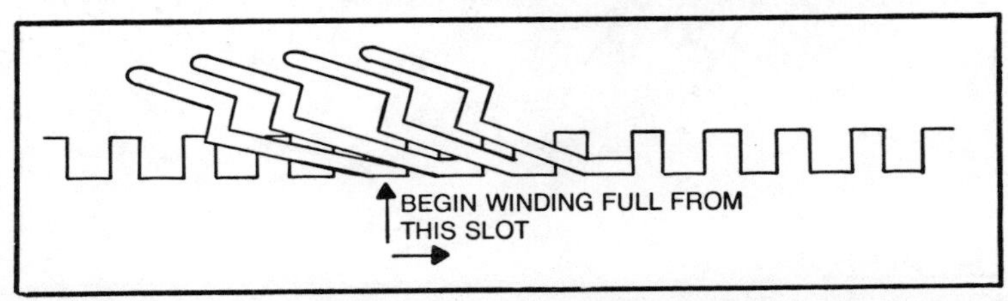


Fig. 11-10. If there are two coils per pole, 4 coils must be left out to finish the winding.

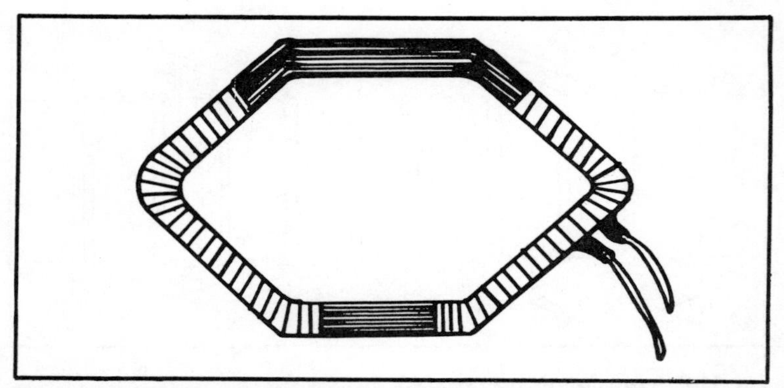

Fig. 11-11. A flat diamond coil. Flat diamond mush coils, as shown here, are only partially taped to facilitate insertion in slots.

time. By leaving a little more tape off to expose the bends, it is easier to manipulate the coil so it can be inserted.

To actually lay the winding, first the slots are lined with plastic sleeves which serve to keep the wires from contacting the metal. Next, a cloth or plastic slider is inserted into the slot with its ends extending outward. Starting on one side of the core, the winder carefully angles the coil in from one end to the other, using the slider as a "lubricant" to allow the wires to move into the slots without catching. See Fig. 11-12. Once the whole side of the coil is in place, it is centered and ready for insertion into the other side.

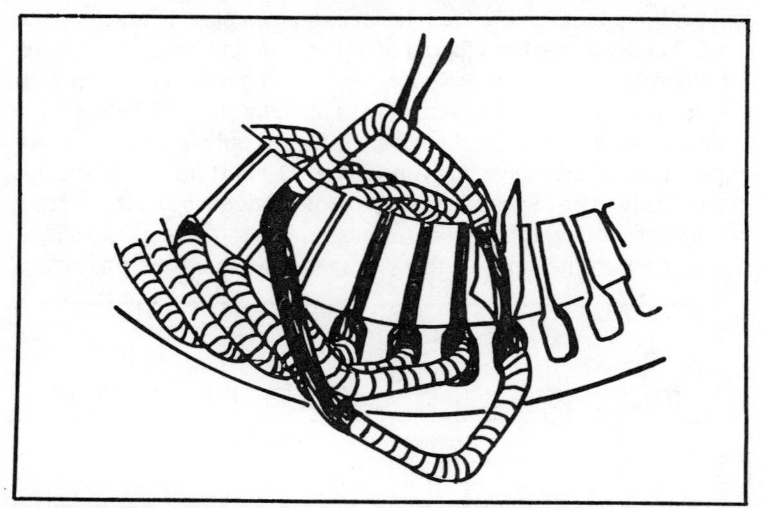

Fig. 11-12. The bottom side is flattened to enter the slot while the top side is threaded in, one or two wires at a time.

While the slider aids in the insertion of the coil, its purpose afterwards is sometimes questionable. Years ago, the insulation used on magnet wire was of a lower grade and temperature rise, that is, it was only allowed to become so hot, perhaps 100° C. Today, higher temperatures are permitted, as technology has produced better insulations which will hold up to more heat. The result is a smaller size motor for the same horsepower rating as years ago. Because of this, the stator plays a key role in conducting heat away from the windings. A good, tight fit between the windings inside the slots conducts heat away readily, while too much cloth and plastic acts as an insulator. The best rule to follow is to leave the slider in if there was one there in the old motor, and take it out if there was none.

The top side of the coil can be swung down and threaded into the slot. If the slot is very narrow, it will take some doing to get it in, if it is a wide slot, the task will be relatively easy. Once the coil is finally in place, it is taped up to the slot like the bottom side. The slot holding the bottom of the coil must be prepared to receive the second coil that will rest on top of it. To do this, the cloth slider (if there is one present) is pulled up in the slot so the wires jam against the opening and stop. See Fig. 11-13 for an explanation of this. The cloth is then trimmed flush with the slot and the coil is pushed back down to the bottom. The cloth ends are folded down and a fiber or wooden stick is placed in the slot to cover the coil. This is to separate the two layers and to protect against shorts. The coil that is to go in the top of the slot is placed in the same fashion as the first. A cloth slider may or may not be prepared just as before, and the coil is inserted. The final wedge is pushed in at the top of he slot, and the coils are finished.

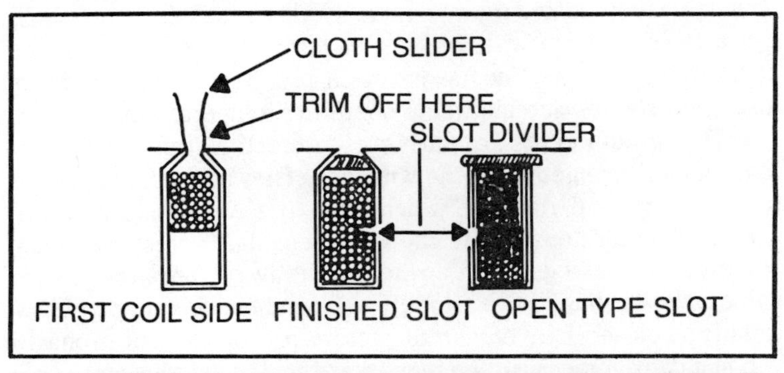


Fig. 11-13. Winding trouble can usually be traced to improper insulation. The proper method of insulating slots is shown here.

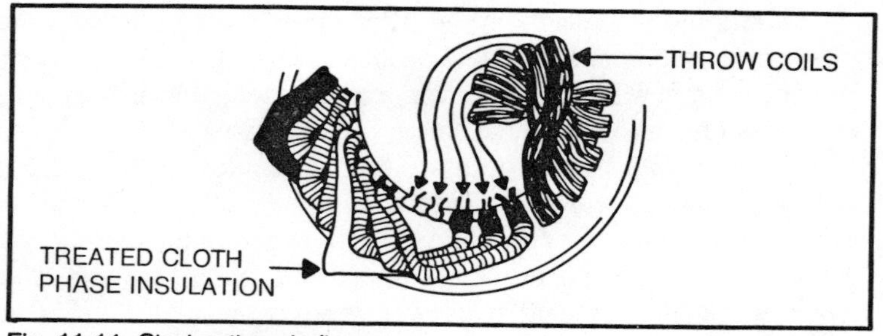


Fig. 11-14. Closing the winding.

As the winding is finished, the coils left out of the slots at the beginning are swung down back over the end ones, as shown in Fig. 11-14. When counting out the pole groups, it is desirable to place plastic or cloth triangles between the coils as shown, to protect against shorts and to make it easier to identify the phases when the motor is connected. The coils are wired together to form the appropriate number of poles and the leads are brought out. Again, there should be a nearly equal number of coils per phase.

## TAPING WINDINGS

The business of taping is rarely done at the factories when the motors are assembled, particularly in smaller motors. The tape does give an added amount of protection to the coils against scrapes, moisture, and dirt, if these elements can enter the motor. The most important thing to remember, however, is that any tape or cloth insulation used when rewinding the motor *must* be able to withstand the operating temperature of the winding. Large, old motors ran fairly cool, the windings never allowed to exceed 105° or so. The newer, smaller motors run considerably hotter, so it is important to use materials which will not melt or burn up on the coils in the course of normal use. See the end of Chapter 3 for an explanation of insulation classes and allowable operating temperatures.

Drip-proof motors are fairly well protected against the problems of moisture getting in the windings. They are still very prone to dust, dirt, and rodents, which love to sleep inside a warm, sheltered motor housing and chew on the coils. Protector screens covering the openings will keep these pests away from an otherwise vulnerable winding. TEFC (totally enclosed fan cooled) or TENV (totally enclosed nonventilated) motors are much less prone to winding damage because the motors are sealed and therefore protected against the effects of foreign materials entering the housing.

An increasingly popular winding type is the encapsulated winding, which is not only difficult to remove but difficult to rewind as well. An encapsulated winding may be a standard one or two layer winding which is covered with a solid mass of epoxy resins outside the slots. To produce this winding at the factory, an ordinary winding (without tape) is fitted with a mold and then injected with a hard setting resin which completely covers all the wires so there is virtually no exposed segments, aside from minute spaces where the molded form ends and the laminations begin.

The first problem with an encapsulated winding is that it is impossible to see what style of lacing is used for the coils, since they are hidden from view. An encapsulated winding has to be chiseled and pried apart to reveal the internal structure. To wind the coils is a simple matter, but to re-encapsulate it is not. Lacking a proper mold, which is a difficult item to construct, mixed epoxy can be packed around the winding by using waxed paper and some sort of putty knife, but this should be attempted *only* after the winding has been tested and verified to be good.

An encapsulated winding is a great way to obtain longer life from a motor than taping. It is also cheaper commercially to encapsulate a winding in an open motor than to produce an enclosed motor with a standard winding. With the sealing, the winding has no exposed portions which could become damaged, so such a motor may be treated more severely than a plain or taped winding. It is not recommended to encapsulate just any motor winding, however, as this may have adverse effects on the cooling if the insulation or the airflow inside the motor is not designed to handle the heat build up which naturally occurs within the epoxy casting. An encapsulated winding is shown in Fig. 11-15.

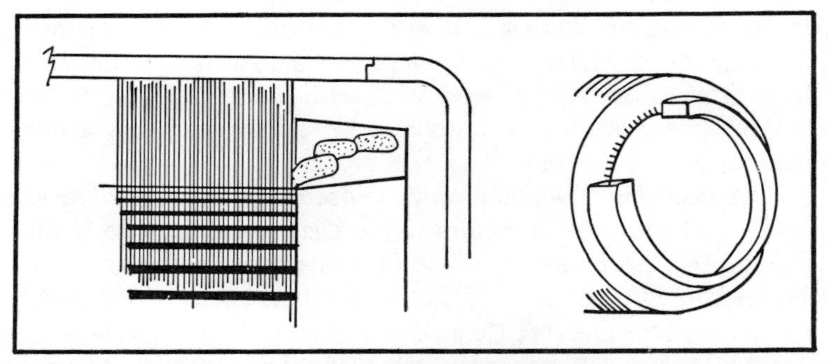

Fig. 11-15. A cross section of an encapsulated winding, showing the molded form surrounding the exposed coils.

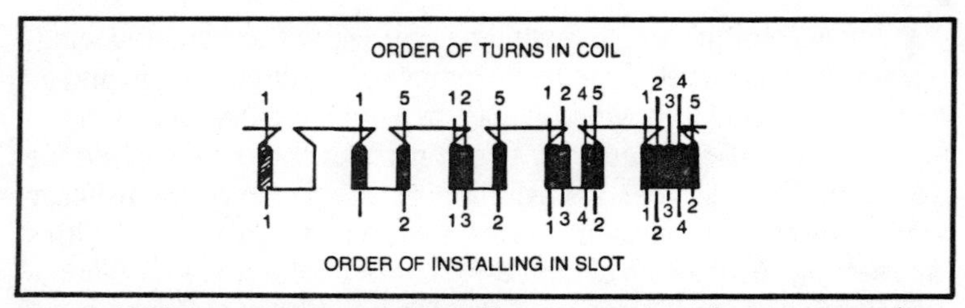


Fig. 11-16. Correct sequence for installing copper straps in narrow openings.

If a motor is particularly large, slow speed, or designed for low voltages, it may be wound with copper straps rather than wires. The copper straps can carry larger currents than a comparable wire wound motor because of their larger cross sectional area. These windings can be either of the lap type or the wave type. The lap type is nearly identical to a pulled diamond winding. These bars must be preinsulated to prevent them from shorting and grounding. This can be done with tape or an enamel covering. If the slot is wide open at the top, and coils can be formed and placed in as a whole. If the slots are semi-enclosed, one strap must be fed in at a time, starting with the outermost and finishing with those which will rest in the center. Figure 11-16 show the correct sequence for laying the copper straps. This is not as difficult as it may seem since the straps are few in number. The only difficulty is in shaping the straps, particularly if they are large and thick. If the windings are of a two layer type, the top and bottom layers are separated by a strip of insulating material just as with the wire wound motors.

The wave winding is shown in Fig. 11-17, along with the lap winding. The only difference between the two is where the end connections are made, either at the center or at spaced points. To decipher either in an old winding, simply note how many conductors per slot and per coil there are, as well as the coil span. Most motors which use these windings are too large for the small shop, and the copper straps are difficult to work with, so the home rewinder or the small motor serviceman will probably never have to worry about rewinding one of these.

One last type of winding which is used mainly on slow speed alternators instead of in motors is the chain winding. The chain winding, like the basket winding, is a one layer version, but it requires two different coil forms to produce the shapes. The winding resembles the links of a chain, hence its name. The coils are well separated at the ends, so there is little chance for shorts, but it takes more time to wind the coils properly than with previous types of windings. Figure 11-18 shows the appearance of a one coil per pole chain winding. The main requirements for a chain winding is that the total number of slots must be even and divisible by two times the number of slots inside each coil. Example: in 11-18 there are 2 slots inside of each coil, so 2×2=4. A stator with 8, 12, 16 etc. slots would be okay.

The main reason why this winding is rarely used on motors is the great number of poles it forms since the size of each coil must be kept small. Induction motors are rarely built with synchronous speeds lower than 514 rpm, hence the chain winding is seldom seen.

The coils are taped prior to their insertion just as before. When starting, however, one or two coils need not be left out on one side to permit the closure of the winding at the end. This type of winding is very easy to lay once it is started, but a different technique is used. Instead of winding the coils consecutively, they are usually laid in two groups; all those which curve downward are placed in first, and then those which are straight are inserted next.

To make the straight coils, a winding machine can be used like the one described in Chapter 3. To make the bent coils, a wooden block is shaped to the general form shown in Fig. 11-19. Holes are drilled into it to accommodate pegs to hold the wire. Once the correct number of turns are placed on the block, the pegs are pulled out and the finished coil is removed. As an alternative, straight coils can be bent down at the ends to the correct shape.

The main problem encountered with the winding of all types of coils is making them too small to fit in the core. Copper coils do not like to be stretched and it is often difficult or impossible to fit a too small coil in the slots where it was meant to go. Another similar case is with two pole windings where the span of each coil is large. The straight distance between the slots in which the coil is inserted

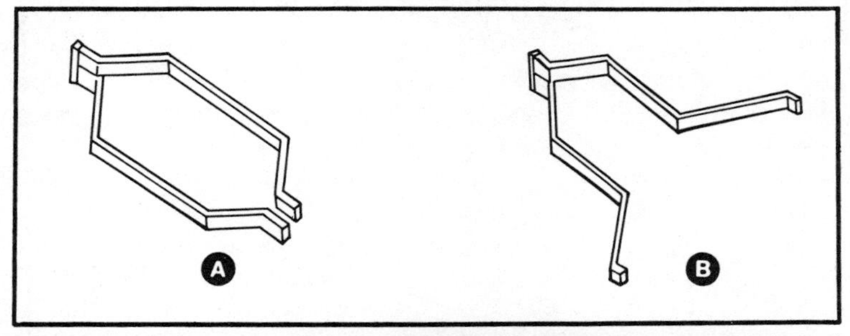


Fig. 11-7. A, a lap winding, B, a wave winding. These may be one solid piece of copper, as shown above, clamped at the ends, or they may be wound with several smaller strands.

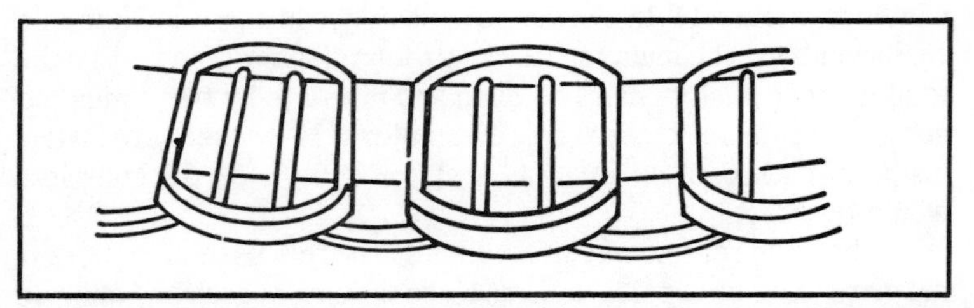

Fig. 11-18. A three slot per pole, one coil per pole chain winding. This same winding could also have been made using two concentric coils per pole, thus giving six slots per pole instead of three.

is not the same as the actual width of the coil, since the curving of the stator shrinks the distance between the coil sides. As a result, a too small coil often sticks up in the path of the rotor. See Fig. 11-20.

The way to get around these problems is to make the coils slightly larger than the anticipated size. If they are too large, however, there will not be enough room for the extra length inside the frame, and the results will be sloppy. The best bet for the beginner is to make up three or four coils and try them in the stator before proceeding to wind the whole group.

Even though the original manufacturer may not place any insulating sheets between the coils of each phase, it is advisable to do so. The cloth or plastic "triangles" mentioned throughout this section give added protection where one phase ends and the next phase begins. Since coils within one group have the same current flowing through them at the same time, there is virtually no potential difference between them, but coils of different phases do. This is why extra insulation is a good precaution against shorts. Many rewinders chalk the slots in which the beginning of each phase

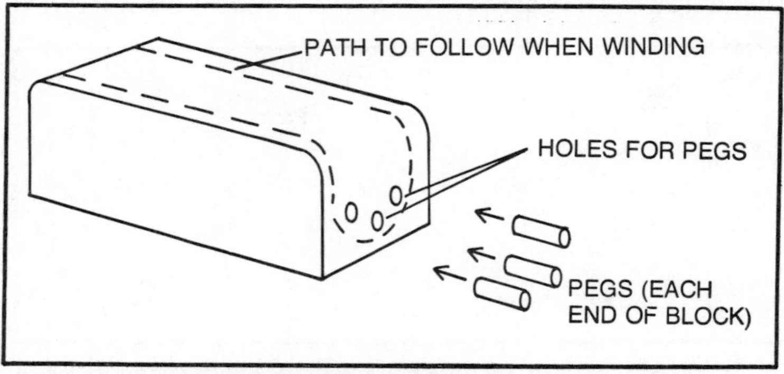


Fig. 11-19. A block for winding chain coils. With this simple form it is easy to speed wind coils of the correct shape.

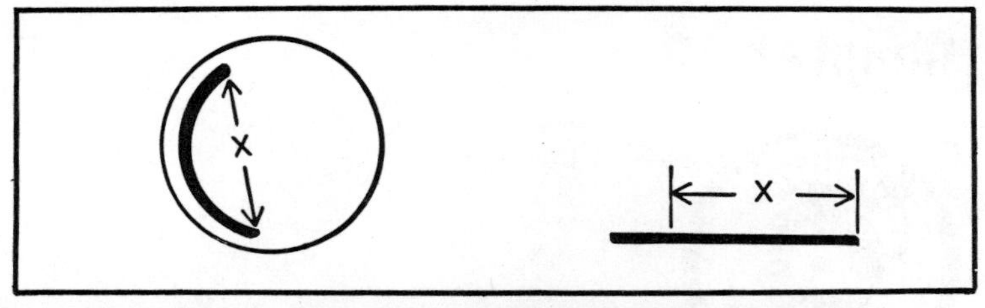


Fig. 11-20. If coil is rolled out flat, it is much longer than distance x.

group of coils will start. This gives an accurate reminder of the places where it is necessary to double the insulation.

As mentioned before, the number of coils in each pole may not be identical, but the overall number of coils in each phase should be nearly equal, for proper phase balance. In a uniform grouping, all groups fill an equal number of slots, and contain an equal number of coils. In the case of alternate grouping (2-3-2-3) every other group consists of an equal number of slots and coils. Irregular grouping is sometimes necessary because of special conditions, and in such cases there is no apparent system of uniformity in the number of slots per group.

If the coils are to be taped, they must be well taped. The tape, which should be able to withstand the heat generated by the winding, must be applied very tightly and anchored in place with quick drying dope, or else it may spring or stretch back and leave an unattached strip. If the slot insulation extends beyond the slots, the tape may be wrapped over this to form a closed bond. Again, the taping must be able to withstand the baking and dipping without dissolving and becoming useless.

All of the coils must clear the rotor and the housing by at least ⅛ inch. If they are too close to the rotor, they will be damaged when the motor runs, if they touch the frame, they can ground it in certain places. The coils can be adjusted by squeezing or tapping gently with a rubber mallet and a fiber drift. The winding must then be tested electrically by the methods described in Chapter 5 before any baking or dipping is done. This insures that any faults can be corrected before the winding is sealed.

# Chapter 12

# Capacitor Motors

The capacitor motor in all its forms has become extremely popular over the other types of single phase motors in the past decades. Its ruggedness and supreme running characteristics have given the capacitor motor a definite place in pumps, compressors, and fans. Home refrigerators and air conditioning units have created a demand for a quiet, high-torque motor which would resist burnouts and create no rf interference, which is a great objection to commutator motors. Much of the credit for the design and improvement of the capacitor motor must be given to Professor Benjamin F. Bailey at the Department of Engineering Research at the University of Michigan. Experiments with motors of this type between the years 1923 and 1929 were financed by the Detroit Edison Company in the interest of the entire electrical industry.

As said before, the problem with a pure single phase motor is that the single phase winding does not produce any revolving flux. The magnetic poles generated by the primary do not move in a circular pattern about the core. They simply jump back and forth across the stator, each pole always being balanced by an opposite pole of equal strength on the opposite side of the winding. Figure 12-1 shows the movement of flux in a polyphase motor as well as a 2 and 4 pole single phase machine. Note the difference.

A rotor inside the polyphase stator will accelerate in the direction of the flux as soon as the power is applied. A rotor in one of the other two windings pictured would only vibrate from side to side. To start the single phase motor, some form of rotating flux must be formed to give the effect of a rotating magnetic field. If a rotor were placed in a plain single phase winding and spun by hand, it would accelerate to full speed on its own and run with the general

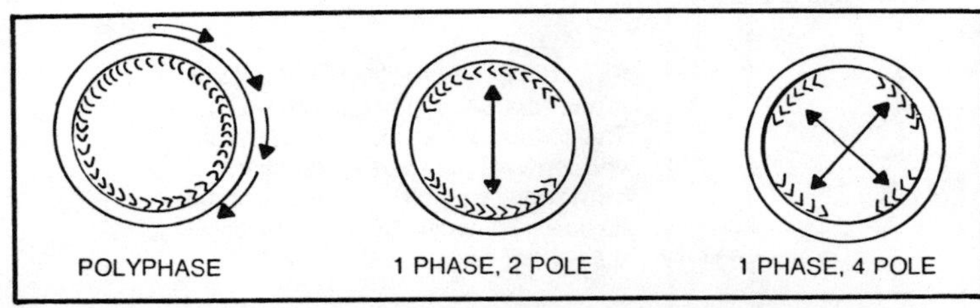


Fig. 12-1. Movement of flux in both single phase and polyphase windings.

characteristics as a two or three phase motor. A manual starting motor, however, is obviously impractical and somewhat dangerous in certain applications.

Many methods have been devised to make the single phase motor self starting. Repulsion and repulsion-start induction-run are two methods used but both require a wound rotor, brushes, and other auxiliaries. Some of these are offshoots of the two phase motor principle. In Fig. 12-2, a two phase winding is shown along with the currents that flow in each leg. They are 90° out of phase with each other. This is what causes the rotating flux. While one leg is beginning to drop another pole reaches maximum strength in a different portion of the motor. This pulls the rotor to one side.

A two phase motor can be operated on a single phase supply if "something" is inserted into one of the phases which will cause the current to lag behind the main flow. A resistor, an inductor or a capacitor may be used. The key to successful operation is to make the phases as close to the two phase current (i.e., 90° apart) as possible. Figure 12-3 shows a two phase motor equipped for single phase operation with a capacitor.

## SPLIT-PHASE MOTORS

The idea of resistance to produce a displaced flux has led to the development of the split-phase motor, a very inexpensive unit

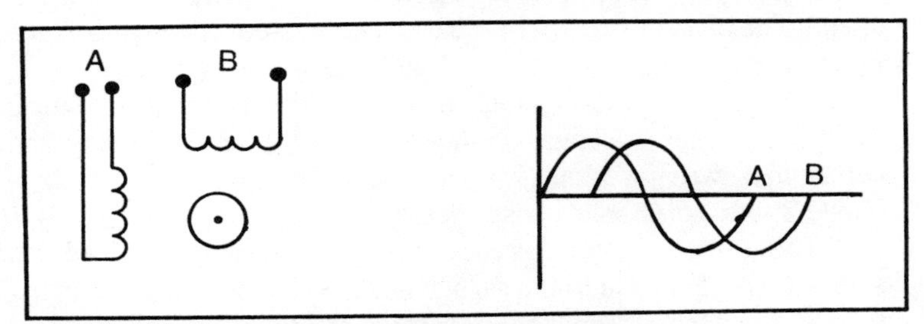


Fig. 12-2. Windings and current in a two phase motor.

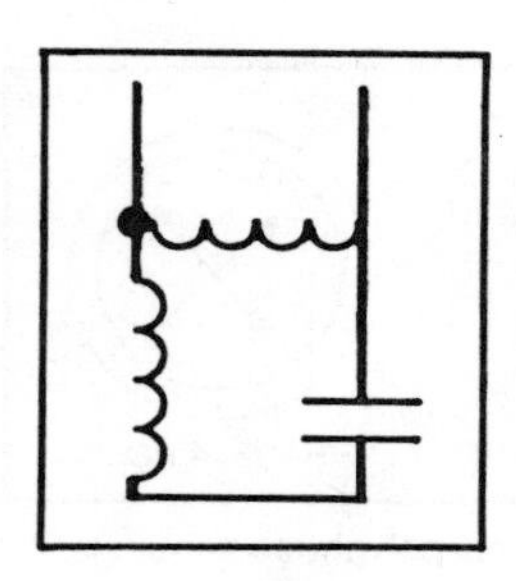

Fig. 12-3. Converting a 2 phase motor for use on a single phase line. The value of the capacitor must be chosen to give the best compromise between starting torque and running torque. To work best for one requirement hampers the other.

which will provide moderate starting torque. The system works as follows: A motor is built with a winding similar to that shown in Fig. 12-1 (the 2 or 4 pole). A second winding is then laid between the coils of the first. The winding is either of the same weight and size as the main winding, or it is made of a finer gauge wire with fewer turns. The finer gauge winding has a much higher resistance than the main winding. If the same size wire is used, a resistor is inserted into the winding to give the same effect as the former. When current is applied, the high resistance winding produces a lagging magnetic field which causes the motor to start. The two schemes are shown in Figs. 12-4 and 12-5.

The effect of the resistance in the starting circuit is not as efficient as that of a capacitor. Why? If a capacitor can displace the phases by 90°, the resistor may only do the same by about 25° or 30°. A smaller displacement does not give the "leverage" which the two phase or the capacitor motor does. Thus, the starting torque is relatively low. Also, when running, the secondary winding does not improve any of the characteristics which are required for proper operation. In fact, the secondary winding, being of high resistance, heats so rapidly that it would burn up (i.e., melt the insulation off the wires) in only a few seconds were it left connected to the line. Because of this, a centrifugal disconnect switch is placed on the rotor which automatically shuts off power to the secondary winding when the motor reaches two-thirds to three-quarters of its full speed. The centrifugal switch is basically a system of weights which fly outward a small distance and pull a plate away from a set of contacts inside the motor's end cover. Breaking the contacts opens the circuit to the secondary winding. Figure 12-6 shows how the centrifugal switch works, as well as the orientation of the windings inside a four pole, split phase motor.

The split phase motor is easy to build, but it cannot be used on a slow starting load due to the reasons mentioned before. Its starting torque is low, and it tends to draw large starting currents. For this

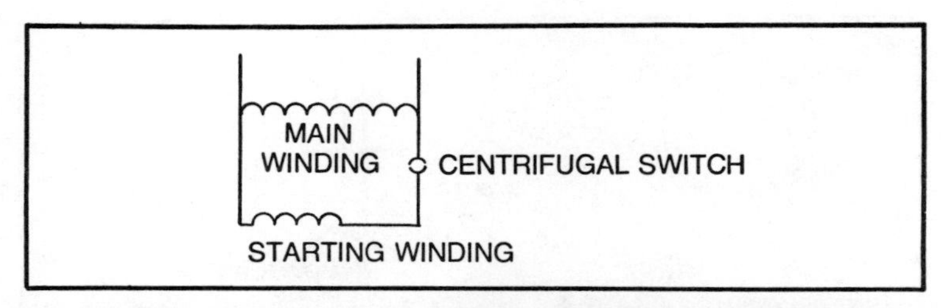


Fig. 12-4. The split phase motor. A high resistance winding is connected with a switch to the main winding.

reason split phase motors are usually smaller than 1 horsepower. Even so, hundreds of thousands of split phase motors are built each year.

Using one of the dynamometers described earlier to measure the torque of a typical split phase motor, one might be able to plot a graph similar to that shown in Fig. 12-7. Notice the point where the switch opens. Also, see how the main winding only supplies a significant torque when the relative movement between it and the rotor is high.

Capacitors also produce a current lag, but to a greater extent than resistors do. If a capacitor is connected to a starting winding with a centrifugal switch, there will be a strong current lag in the secondary winding which will start the motor with good torque and a minimum of starting current. It was known that capacitors would do this many years ago, but until electrolytic capacitors could be manufactured in small sizes at low prices, it was impractical to use a capacitor for starting which was as large or larger than the motor itself. The essential difference in the split phase motor shown in Fig. 12-4 and that illustrated in Fig. 12-8 is that in the former, the higher resistance of the starting winding causes an out of phase relation between the current in starting winding and the current in the main winding, while in the capacitor motor shown in Fig. 12-8 an

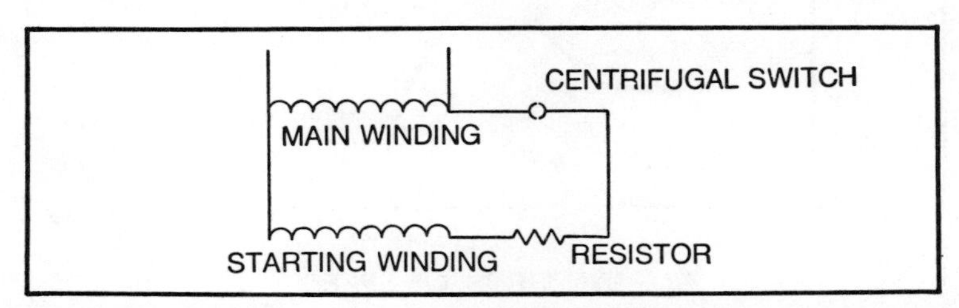


Fig. 12-5. A variation of the above. A resistor is used to add resistance to the secondary winding. This method is almost never used today.

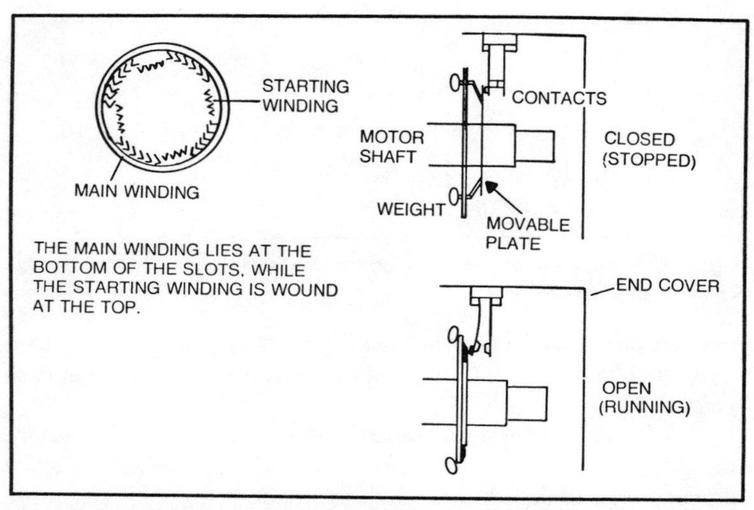


Fig. 12-6. The centrifugal switch. As the motor accelerates, small weights are moved outward and relieve pressure on a spring loaded contact. This sketch is only one simple type of switch. There are other designs of weights and movable plates.

external capacitor connected in series with the starting winding causes a similar out of phase relationship.

Nearly all single phase motors are of 15 horsepower or less, and the maximum value for the starting capacitors used might range from 1000 μF down to 25 μF for motors of 1/6 horsepower and less. The size chosen does not have to give a true 90° displacement, as long as the displacement is large enough.

If two motors, a split phase and a capacitor start, are of the same horsepower and draw the same starting current, the capacitor

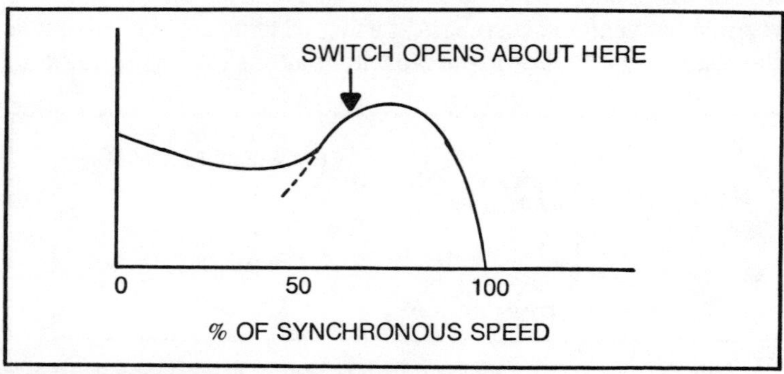


Fig. 12-7. Starting torque of a split phase motor.

motor will produce about double the starting torque than the split phase motor. If the starting torques are designed to be equal, the capacitor motor will draw less current and start cooler than the split phase motor.

The capacitor motor does have a drawback. If the starting capacitor is left in the circuit, it will hamper the running characteristics of the motor. Also, like the split phase motor, the capacitor itself will explode if it becomes overheated due to frequent starting or long starting periods. Thus, a centrifugal switch is used to cut the starting winding out of the circuit as before.

Capacitor start motors do not usually suffer from too long a starting period, however, because of the high torque they develop. In some cases, the capacitor start motor can deliver 3 or even 4 times its running torque while starting. Look back at Fig. 12-7 and note how the torque of the split phase motor drops as it accelerates. The capacitor start motor does not do this. Instead, it increases to a maximum as it accelerates, up to a peak at perhaps 80% of synchronous speed. It then falls to normal full load torque. Because of this increasing torque, capacitor motors can start heavy loads without much strain.

Although the large starting capacitor is too big for operation at normal running speed, a smaller capacitor can be used to improve the efficiency at full load. If a small, high resistance run capacitor is left permanently connected in series with the starting winding, without a centrifugal switch, the result is called a *permanent split capacitor motor*. Since the capacitor is only a few microfarads, it does not produce a great displacement between the starting and running phases and so the starting torque is very low. However, at running speed, the small out of phase current in the secondary winding improves the efficiency and running torque a good extent. For example, a plain induction run motor may draw 500 watts, while

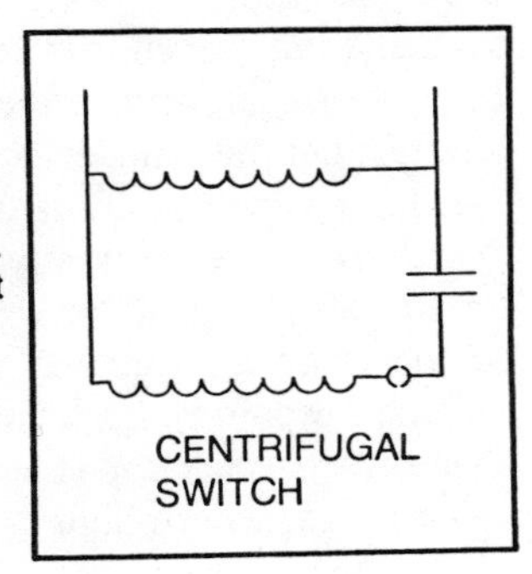


Fig. 12-8. A capacitor start motor. Both windings are equal in weight and turns.

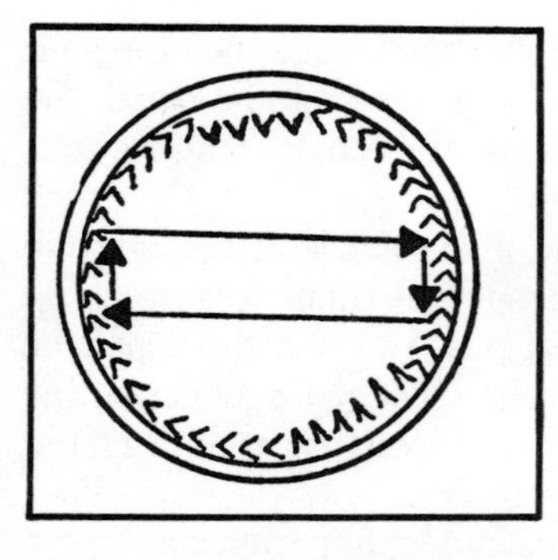

Fig. 12-9. A 2 pole, permanent split capacitor motor. The tiny forces indicated by the short arrows soften the pulsating effect in the pure induction winding.

a permanent split capacitor motor of the same horsepower may draw 400 watts. In addition, the capacitor motor runs much quieter.

To understand this, refer back to the flux inside a single phase motor in Fig. 12-1. Note how the arrows jump back and forth inside the core. This flux movement causes a great deal of pulsation inside the motor, sort of like shaking a rubber ball up and down inside a jar. It is this reason that most single phase motors are made with resilient mounts. Figure 12-9 shows the flux movement inside a permanent split capacitor motor. The two displaced forces increase the number of jumps to 4, the effect of the ball and jar model being bounced at 4 instead of two positions. This decreases the total force (the length of the large arrows) of the main pulses and hence reduces vibration somewhat. The result is a quieter running motor.

Because the starting torque of the permanent split capacitor motor is so low, it is used primarily on fan loads, where it performs quite well. To increase the starting torque of this motor somewhat, sometimes a split phase version of this motor is used. Instead of leaving the capacitor permanently connected to the starting winding, a centrifugal switch is placed in the circuit and a high resistance winding is used. The centrifugal switch, now a double throw design, allows the motor to start like a plain split phase motor and then place the run capacitor in series with it at the proper speed. Figure 12-10 shows this arrangement. This motor has good running characteristics but only moderate starting performance. The problem of the one capacitor motor is that the one capacitor does not allow the motor to operate efficiently under both starting and running characteristics. If the capacitor was of the right value for starting it would be too large for running, while if the capacity was suitable for running it would be inadequate for starting. Nevertheless, the split phase start, capacitor run motor is extremely useful for furnace applications, where high starting torque is not needed, but a low cost, quiet motor is ideal. The energy saving of the capacitor run motor over the plain induction system is desirable whenever items are used frequently and energy cost is a consideration.

The characteristics of the capacitor are the most important factors in the starting and running efficiency of the motor. To obtain the best results, the capacitor should be large for starting and then somehow reduced gradually as the rotor speed increases. The highest efficiency could be obtained by varying the capacity to suit the changes in load, but this would require a complicated system of taps and switches. On the other hand, the use of a fixed capacitor suitable for the running period only will provide about 50% of the running torque for starting. This again is the split capacitor motor, which is often called a compromise between the start and run capacitor motors.

## SINGLE-CAPACITOR MOTORS

The single capacitor start motor, called a capacitor-start induction-run motor, does not approach the usual two phase efficiency when running. Therefore, a superior motor with one capacitor for starting and one capacitor for running is often used. This gives high efficiency and power in both modes and the result is a remarkable motor. Figures 12-11 through 12-13 show various ways of building such a motor. The difference is in how the centrifugal switch operates the windings and how the capacitors are placed in the circuit. The starting capacitor is a high value electroly-

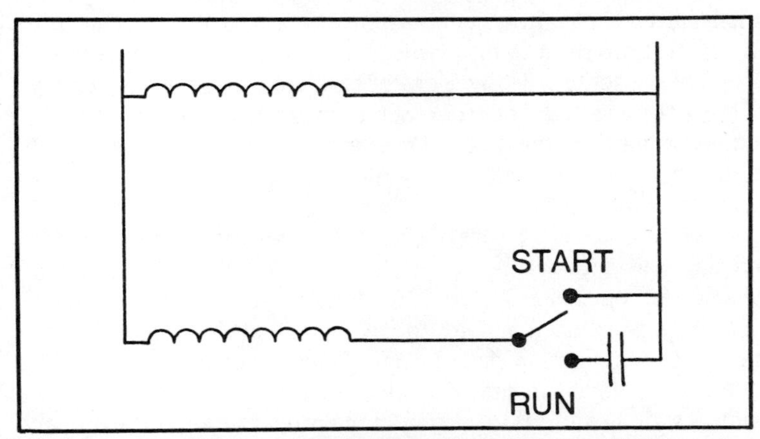


Fig. 12-10. The split phase start, capacitor run motor. The small capacitor is switched into the circuit to keep the starting winding from burning out and to increase the running efficiency.

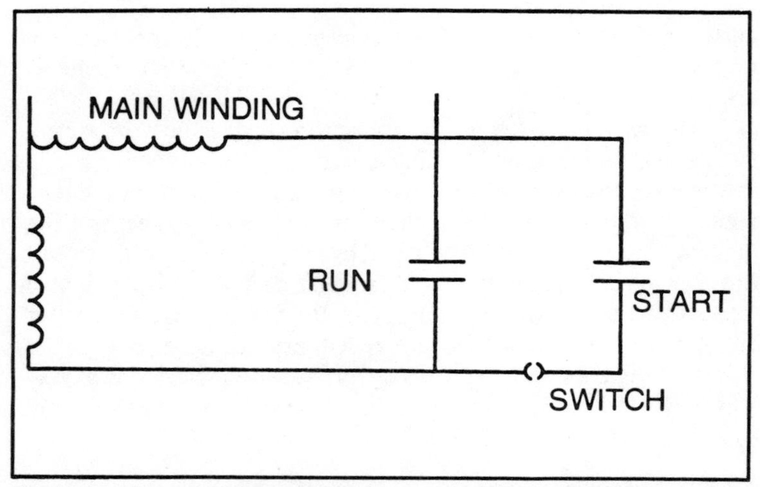

Fig. 12-11. This system leaves the run capacitor permanently connected.

tic type, while the run capacitor is of a high resistance, low value oil type. The running capacitor may be left in the circuit permanently or it may be switched in.

The capacitor start and run motor is generally considered to be the best of the single phase motors due to its similarity to two phase operation in both the running and the starting modes. It is, however, the most expensive of the single phase motors and this sometimes deters the user from making the initial purchase. Because the high torque of this motor is not needed for some applications, such as small fans, it is not always necessary to use a capacitor start and run motor. Therefore, it is mainly used with heavy duty devices.

The capacitors, be there one or two, are usually housed on top of the motor in a metal cylinder. Sometimes a single run capacitor is simply strapped to the motor if the capacitor body itself is made of metal and sealed. Figure 12-14 shows various motor and capacitor housing configurations.

As said, two capacitors of different values can be used, one for starting and accelerating to maybe 75% of full speed, and the second, for operation above 75%. What if three, four, or more capacitors were used and set to cut in and out over smaller segments in the operating range? The efficiency and power would be increased a slight amount, but the difficulty in adjusting the centrifugal switches to cut in and out in the proper order at the proper times would offset the performance. A small measure of efficiency is sacrificed, therefore, in the interest of simplification.

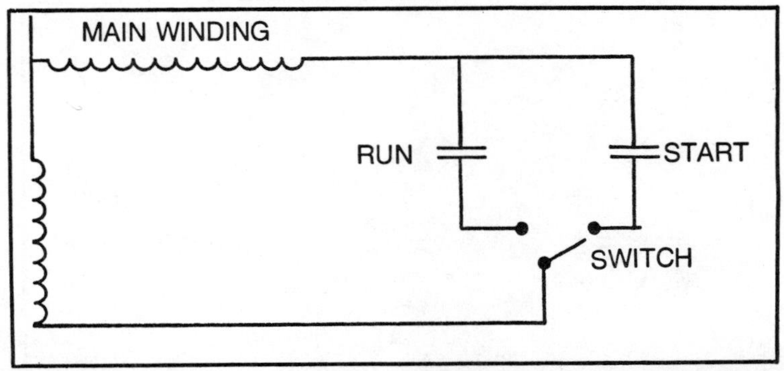


Fig. 12-12. Here the run capacitor is switched in after the start capacitor is switched out.

## MULTI-CAPACITOR MOTORS

Figure 12-15 shows a multi-capacitor motor. At starting, all of the capacitors are connected and they are then cut out one at a time as the motor accelerates until only one is left in the circuit. The switches could be precision centrifugal disconnects, or *current relays*. Current relays are coils connected in series with the supply which hold a set of contacts closed while the current flowing in them is sufficient to cause an electromagnet to pull them shut. If the current is reduced, the relay opens up and thus cuts out the capacitor. By using the proper relays, ones which had different current requirements to open and close, it would be possible to eliminate the centrifugal switches.

The relays themselves would have to precisely set in order to switch the motor for the best performance, and the cost would not

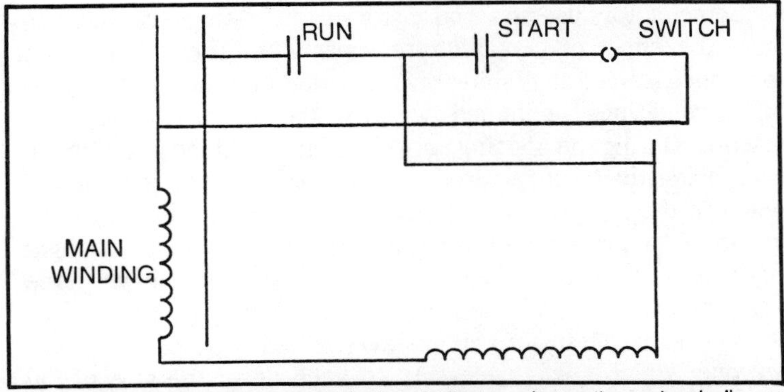


Fig. 12-13. This system center feeds the windings and uses the main winding as an inductor in series with the starting winding.

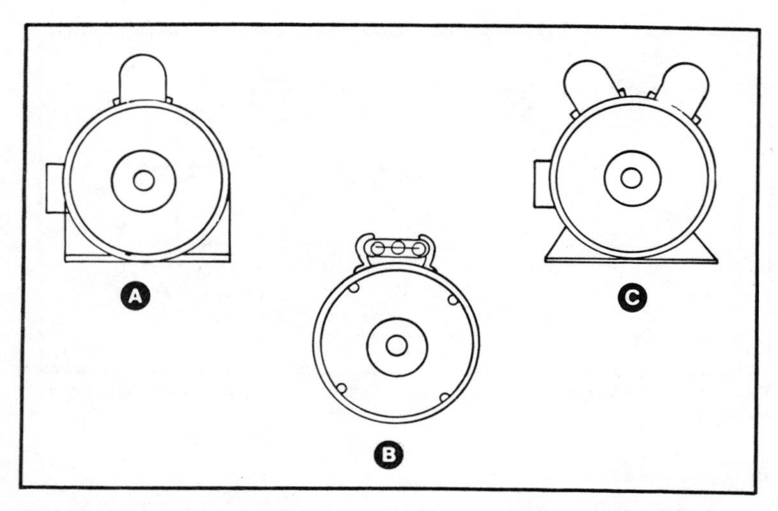


Fig. 12-14. A, shows how a single capacitor may be housed on a motor. B, shows two capacitors. C, illustrates how some capacitors are strapped to motor frames without being completely covered.

be any less than with centrifugal switches, so it is really a matter of the size of the motor which determines what system is used. Centrifugal switch contacts are small and cannot carry large currents, while big relays which can be located remotely from the motor and need not be confined to the small space inside the motor. If such a motor were running at full load and then the load was increased, the motor would slow and one or several of the capacitors would be reconnected by the switches to provide added torque. This should speed the motor back up, and after the overload had passed, it would continue to run as before.

Some motors use an autotransformer and a single capacitor to obtain the effect of two different capacitors. The value of the capacitance is fixed at all times and the variation to suit starting and running conditions is obtained by varying the voltage applied to the capacitor. During the starting period, a higher voltage is applied to the capacitor than that required for running. The effect is about the same as if the capacitance was changed, as in a capacitor start and run motor, but the transformer reduces the efficiency a slight amount by its own internal losses. There are two ways in which the transformer can be switched, one is shown in Fig. 12-16. This method of using a transformer, however, is nearly obsolete and was used only when it was necessary to keep down the size of the capacitor. The system can also be operated manually if a variable

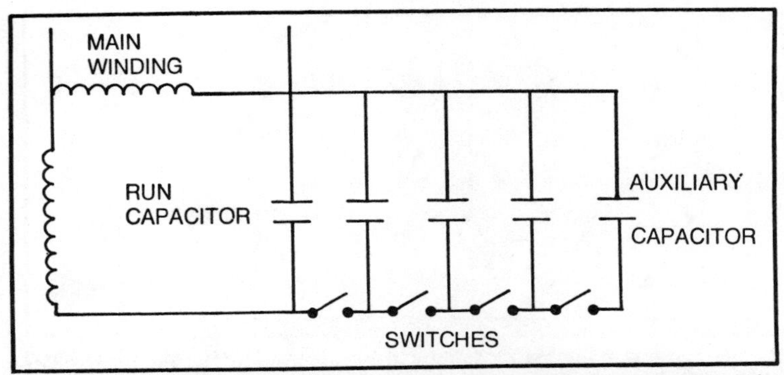


Fig. 12-15. A variable or multi-capacitor motor. Such motors were built mainly for experimental purposes and were never actually used. Today, using more efficient motor design, the extra capacitors are not needed. Higher operating temperatures allow for smaller motors with better power factors and efficiencies than were possible years ago.

autotransformer is used. Here, the operator turns the transformer to a preset value for starting and then reduces the voltage as the speed increases. This is of use only on a slow starting load, where the time to make the changes would be significant. Such a plan is shown in Fig. 12-17.

Now that we have seen how the size of the capacitor affects the starting ability and power factor of the capacitor motor, let us consider the motor windings. The ratio of turns in the two windings as well as the weight of the copper has an important effect on performance. The starting torque of the motor can be influenced considerably by changing the ratio of turns in the windings.

The size of wire used, as well as the number of turns in the main winding are more or less definitely established for a motor of

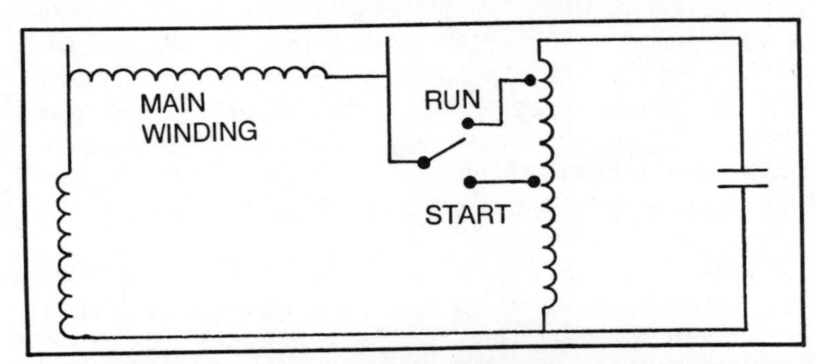


Fig. 12-16. A two tap autotransformer is operated by a switch to vary the voltage to the capacitor.

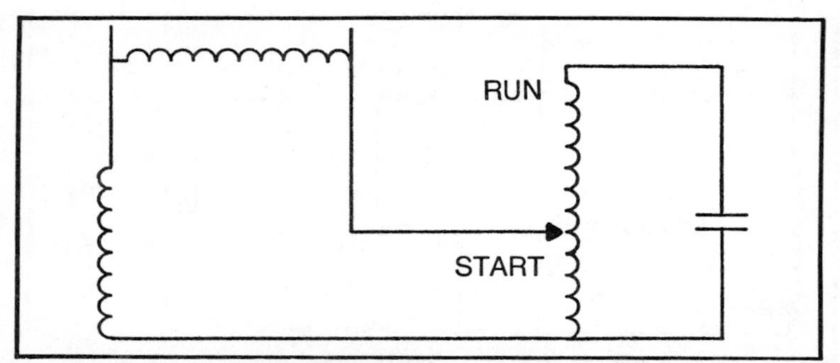

Fig. 12-17. Using a variable transformer to provide greater regulation than that in Fig. 12-16.

any given output. These are computed by the manufacturers to give the best performance with the least amount of adverse side-effects, such as heat, noise, vibration, and high currents. If the number of turns is too large in the main winding the pull-out torque will be too low for many uses. This is because a large winding has a high self inductance which does not allow a high current to flow through it when a lot of power is needed, such as in starting. If the number of turns is too small, the efficiency and the power factor will be seriously decreased because there is not enough of a magnetic field produced to run the motor properly. For these reasons the original design should be followed closely when rewinding the main phase winding in a capacitor motor.

When replacing a capacitor, the same size should be used or the next largest size if an exact replacement is not available. The capacitor winding, however, can be varied to a great degree so as to obtain differing performances and conditions. If the number of turns in the starting winding is doubled, the wire size reduced to one half the former cross section, and the capacitor replaced with about one fourth the former value, the torque will be reduced approximately 50%. This will also reduce the starting current. It has the effect of putting in a higher voltage winding than the original. If more starting torque is required, the winding can be made with a heavier gauge wire and fewer turns, with a larger capacitor. This will, of course, increase the starting current.

Modern day capacitors for electric motors are almost all of the electrolytic type. They are small and inexpensive, but they suffer from internal losses which lower the overall efficiency of the motor. These capacitors can only be used intermittently for starting, as the high currents will damage them. Paper type capacitors, which were

**Table 12-1. Comparison of Different Types of Motors.**

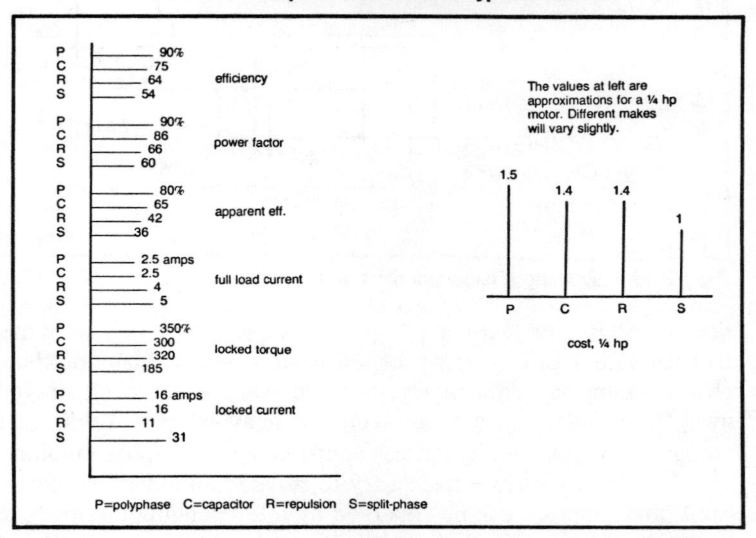


once used can handle larger currents, but they are very bulky. Oil type capacitors used for the run modes are generally rugged and moderately expensive.

So far, this chapter has given the basic rundown on split phase motor development and the various forms of capacitor motors. The points to remember are: the capacitor motor has a higher efficiency than other single phase motors, it has a high power factor, it has excellent starting torque, it is easy to build, it is quiet, and it is reliable. Capacitor motors are more expensive than other types but the advantages outweigh the disadvantages.

Capacitor motors are used frequently in pumps, compressors, air conditioners, conveyors, and in furnaces. Keep in mind that any

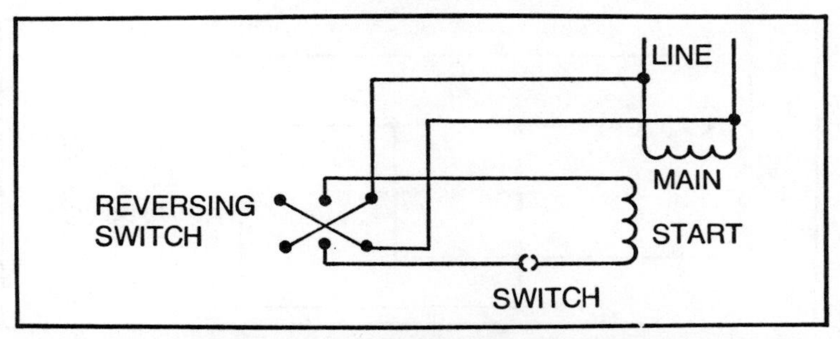


Fig. 12-18. Reversing a split phase motor if all four leads are accessible.

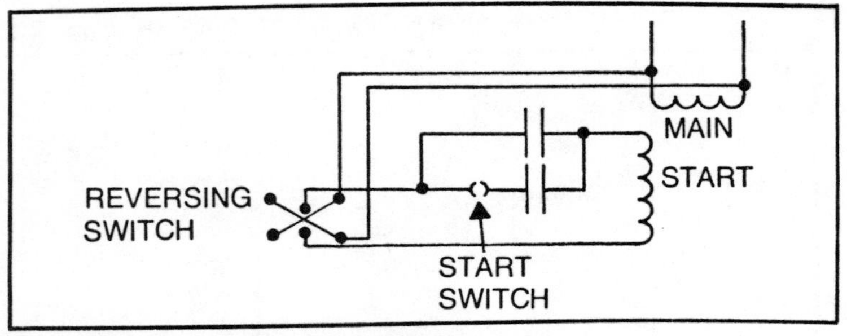

Fig. 12-19. Reversing a capacitor start and run motor.

household device with a pump or compressor (such as a refrigerator) is a prime target for a capacitor motor. Many machine shops employ capacitor motors to run power tools, having strayed away from split phase motors on the heavier jobs. Table 12-1 compares polyphase, capacitor, repulsion, and split-phase motors.

In many cases it is necessary to reverse an induction motor. Split phase motors can be reversed by interchanging the leads to one of the phases. Often, the terminal board inside the box will have a provision for this. While the split phase motor is not instantly reversible, that is, it must slow down to the point where the centrifugal switch cuts back in, the reversing can be accomplished by a dpdt switch, as shown in Fig. 12-18.

Capacitor start motors can be reversed in the same manner. If all the leads are not accessible from the terminal box, the motor must be opened to make the proper connections. Often the terminal board has a system of spade lugs mounted on it which allow the motor to be reversed by simply interchanging two of the lugs. This is exactly what the switch does in Fig. 12-18. By bringing the connections out of the conduit box, switches can be added to allow the motor to be reversed by a simple flip.

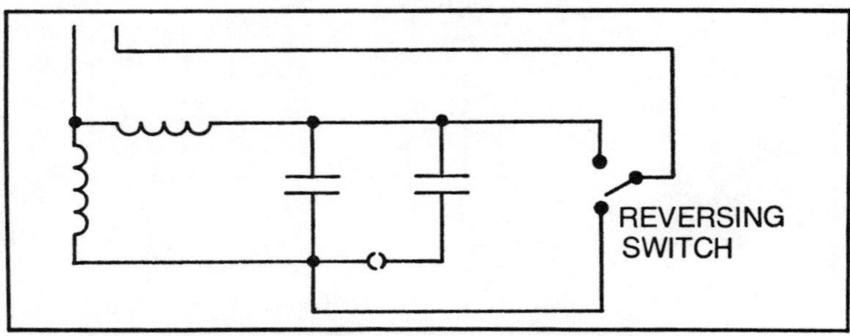

Fig. 12-20. Reversing a capacitor motor with an spdt switch.

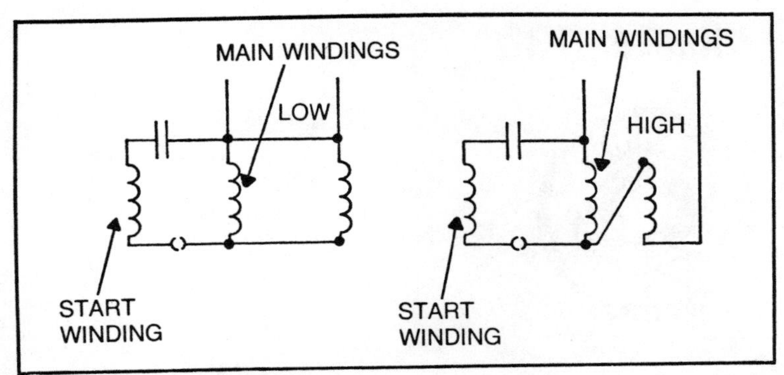


**Fig. 12-21. Dual voltage operation.**

Capacitor run motors are a slightly different situation. The run capacitor, being permanently connected to the winding, will create a displaced phase in the same direction no matter how the rotor is turning. In one direction, it would run normally. In the other, it would tend to slow the motor down. If the value of the run capacitor was such that it provided nearly equal torque from the main winding, the motor would stop. Otherwise, it would simply interfere with the motor's operation. Because of this, *both* the starting and running capacitors must be reversed together if the motor is to run properly. Figure 12-19 shows one way to do this with an dpdt switch.

The other alternative is to reverse the windings, i.e., make the primary the secondary and vice versa. Figure 12-20 shows this type of control with an spdt switch. Study it carefully. Note that if the switch is moved from one position to another, the connection of the run capacitor in relation to the line is changed.

One last note on capacitor motors. Often it is desirable to be able to operate the motor on two different voltages, perhaps 110 and 220 or 220 and 440. If a given coil of wire is rewound with a smaller diameter wire, with twice the number of turns as the original, it will operate the same at double the original voltage, providing that the weight of the new coil is the same as the old.

With a capacitor motor, the coils may be connected in parallel to operate at a lower voltage and in series for operation at a higher voltage. Figure 12-21 shows the typical hookup. Note that the starting phase does not change at all since it is in the circuit for such a brief period that it can handle either voltage.

# Chapter 13

# Reversing Rotation of Ac and Dc Motors

Aside from burned out windings, worn bearings and rough commutators, one of the main causes of having motors brought to the shop is that of having the direction of rotation reserved. Such a task, no matter how simple it may be for the service man, is usually a great mystery to the customer. In most cases the owner of the motor has picked it up to operate some piece of machinery and finds out later that it does not turn in the right direction to suit his need.

Most motors are easy to reverse if the principles involved are understood. Different methods must be used to reverse the rotation of the various types of motors, and a small amount of space will be given here covering the types in most popular use.

First to be considered will be direct-current motors of the types in general use, as the same instructions will do for 6, 12, 32 and 110 volt machines.

Direct current motors are reversed by reversing the direction of current flow through either the field or the armature, but not both. The circuit of an ordinary dc motor of the series type is shown in Fig. 13-1, and that of a shunt connected motor in Fig. 13-2. As it is usually easier to change the field leads at the brush holders, thus reversing the path of the current through the armature, this is the means most often employed. In the figures, the original connections to the brushes are shown as solid lines while the changes made to reverse the rotation of the armature are indicated by broken lines. It is sometimes necessary to lengthen either one or both of the field leads in order to reach the desired brush holders, or to keep the wires from passing too close to the commutator.

## DC MOTORS

On the smaller dc machines, and those having either solid end bells or ones with small inspection holes, it will be necessary to remove the commutator end housing to accomplish the job of reversal. In any case this is usually a small matter and the entire job can generally be completed in a half hour at most. Small universal motors—ac or dc are reversed in the same manner as the series direct-current motor.

## AC MOTORS

Alternating current motors are reversed in several different ways, depending upon the type of motor under consideration. In the case of the common splitphase motor we must get at the junctions of the windings in order to change the direction of rotation, and this means partly dismantling the motor.

The split-phase motor has two distinct windings, the heavy low resistance running winding, and a smaller high resistance starting winding. To change the direction of rotation of the rotor the terminals of either one of the windings must be interchanged. That is, the line terminals of the running winding can be switched in relation to the line, or the two ends of the starting winding can be shifted on the line. Do not make the mistake of changing the terminals of both windings or no change of rotation will result; leave one winding as it was found.

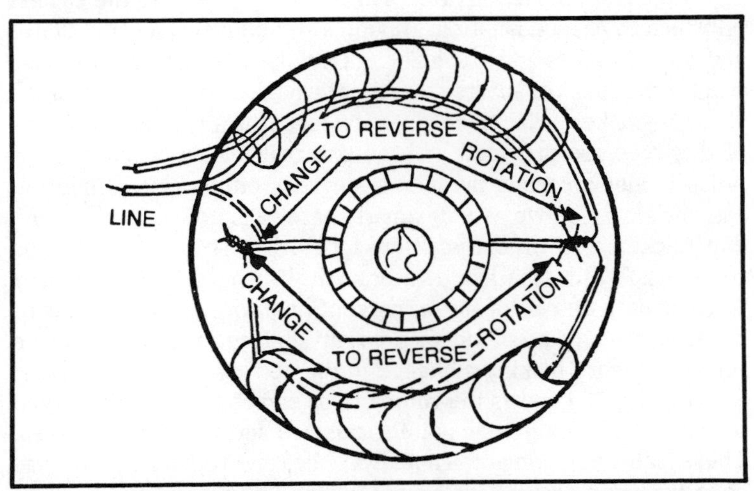


Fig. 13-1. Changes necessary to reverse rotation of a series motor shown by dotted lines.

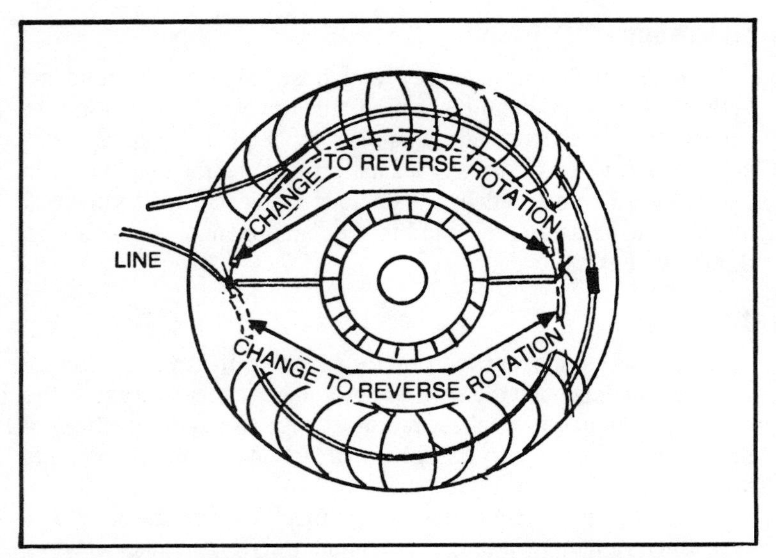


Fig. 13-2. Changes necessary to reverse rotation of a shunt wound motor are shown by outline marks.

Figure 13-3 shows a schematic diagram of a split-phase motor winding. The points marked 1, 2, 3 and 4 are the connections of the two windings to the line. To change rotation interchange 1 and 4, or interchange 2 and 3, but never all four points.

Repulsion-induction motors are perhaps as a class the easiest on which to change rotation. Having a wound rotor, a commutator and set of brushes and brush holders similar to dc machines—save that the brushes are shorted together through the metal of the brush rigging—makes a change in direction mechanically simple. Motors of this type are provided with an external means of shifting the brush holder a certain number of degrees either side of neutral. Narrow slots, known as index marks, are cut into the metal of the end bracket in the proper positions for clock-wise or anti-clockwise rotations. Another mark or pointer on the brush rigging can be made to coincide with one of these marks. Some form of lock screw or clamp is usually provided to prevent shifting, and this must be released before making a change in the setting.

Figure 13-4 shows the commutator end of a motor of this type in which the index marks, pointer and lock screw are plainly seen. These factory markings can not always be relied upon if a motor has been rewound or has had a new commutator installed. A slight change in the winding or position of the commutator on the shaft will cause a shift of the true neutral position of the brushes—a point at which the armature will revolve in neither direction—and consequently will likewise shift the brush positions for best operating conditions. By means of instruments on a test bench, or by means of a starting torque test, the position of the brushes for greatest efficiency in starting can readily be located. New marks can then be made on the motor frame.

It is sometimes desirable to have a reversible motor on such a piece of machinery as a lathe, when the expense of purchasing a true reversing motor is unwarranted. For occasional work of this kind at constant speed, a repulsion-induction motor can be adapted. By bolting stops at each end of the index marks, some form of hand control can be attached to the index pointer, so that a quick shift can be made for rotation in either direction. This type of motor must, however, be allowed to come to a full stop before attempting to run it in the reverse direction, or serious burning or arcing at the commutator will result.

Straight induction motors are easily reversed in most cases by exchanging the end bells so that the rotor can be changed end for end. As a rule motors of this type are built so that either end housing will bolt to either side of the stator housing. By switching the rotor end for end, or by leaving the rotor as it is and turning the stator around, the rotation of the machine will be reversed without any alteration of the windings or connections. Figure 13-5 shows this clearly.

Shading-coil motors can not be reversed by changing leads or terminals, as this type of motor, used mainly in ceiling fans, has but one winding connected to the line. The shading-coils, which are used to make this form of motor self-starting, are placed to one side of the main poles, and would have to be shifted to the opposite side of the main poles to cause a reversal of the direction of rotation. In

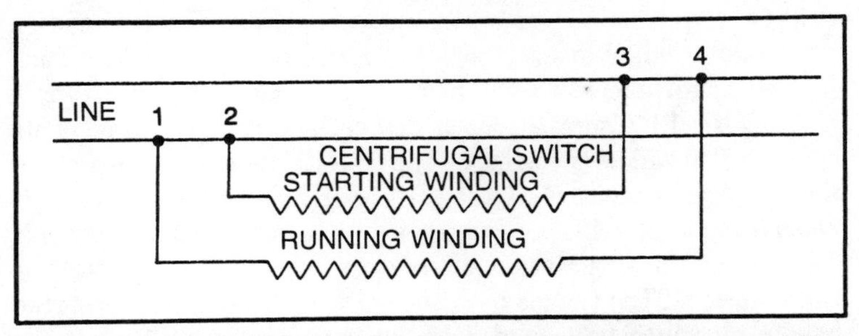


Fig. 13-3. Connections of a split-phase motor. To change rotation reverse either terminals 1 and 4 or 2 and 3 in relation to power line.

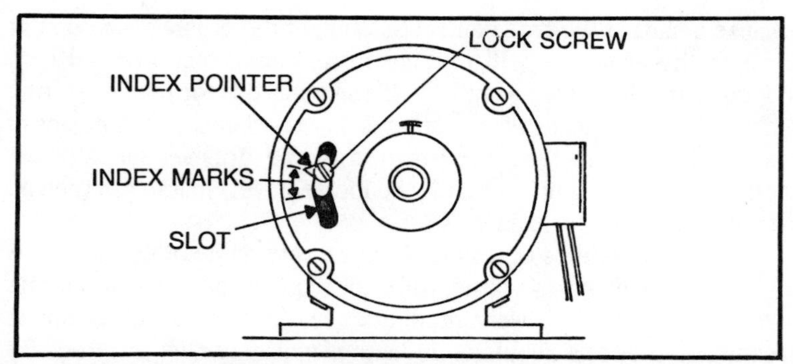

Fig. 13-4. Reversing rotation of a repulsion start-induction run motor. Shifting index pointer to other index mark changes rotation.

most cases this would be a practical impossibility, as it would necessitate a complete rebuilding of the field, and is a method hardly ever used.

The shading or starting coils used in these motors consists of either a circular copper band, or of a short circuited coil of heavy wire. This band or coil inserted in the trailing edge of the pole retards the flow of the magnetic flux in this side of the pole and causes a phase displacement which has a rotating effect on the rotor. These starting coils are insulated from the main winding. See Fig. 13-6.

To change the rotation of a shading-coil motor, reverse the end bells and rotor in relation to the field. If this is impossible because of mechanical restrictions, it may be possible to press out the stator laminations with the windings intact, and press it back into place in a reversed position. In the latter case the rotor, of course, must be left in the original position in the frame.

After rewinding the stator of an induction motor, or before assembling the complete machine, it is often desirable or essential to know in which direction the motor will operate. This information can be determined in advance with the use of a simple shop made tool which, for want of a better name, can be called a "testing rotor."

Figure 13-7 shows such a device, and a description of its construction will be given here. The material needed to construct a testing rotor consists of two discs of fibre, bakelite or hard rubber, and a few feet of bare copper wire in several sizes. This material is formed into the shape of a squirrel cage, with two coils projecting from the ends. The cage is mounted on an axle which also forms the handle by which it is held when in use. Figure 13-7 gives the approximate dimensions of the tool.

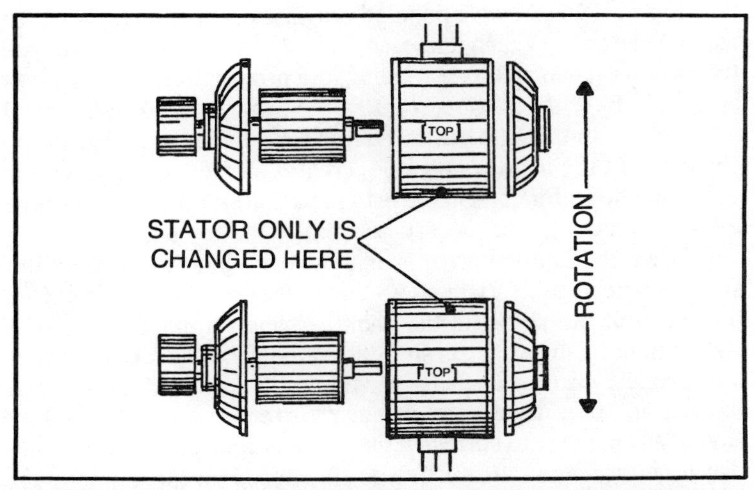


Fig. 13-5. To reverse rotation of an induction motor change rotor end for end, or change sides of stator, but not both.

To discover the direction in which the assembled motor will rotate, connect the stator windings to the line, and with normal current flowing through the coils, hold the testing rotor inside of the stator opening close to the core. Observe the direction in which it tends to turn. This direction will be the same as the rotation of the assembled motor.

Another handy testing device for the motor repair or rewinding shop is an internal growler. Just as an ordinary growler detects faults in most armatures, so will the internal growler locate short circuits inside the field ring or stator. Such a piece of equipment as just mentioned can be made at little or no expense by any electrician, and will be well worth the time and material expended on it.

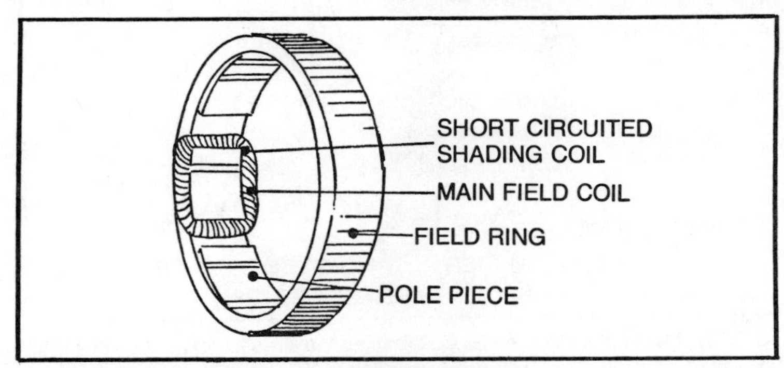


Fig. 13-6. Construction of a shading-coil motor.

Figure 13-8 shows a detailed drawing of the internal growler and the manner in which it is used. In reality this tool is a core type transformer having but one winding, the primary. The primary coil is excited by 110 volt ac current as the growler is passed around inside of the stator. The laminated iron core of the internal growler, shaped to fit the inside of a circle, creates an alternating magnetic field, and the stator coils in the path of this alternating field become for the moment the secondary winding of the testing device.

When the internal growler is moved around the inside of the stator—touching the stator iron—and passes over a short circuited turn or coil, the result is the same as when the secondary of a conventional transformer is short circuited. The short circuiting of the secondary causes an increased flow in the primary of the transformer, and this increase of primary current can be detected by having an ammeter in series on the 110 volt line. At the same time that an increase in current flow is registered on the ammeter, it usually happens that considerable heat is generated in the short circuited section of the secondary of the stator. After a moment or two the defective turn or coil of the stator can be located by feeling over the surface just tested with the bare hand.

Another method of finding shorted coils with the internal growler is as follows: Just as a shorted armature in an armature growler will cause attraction for a strip of metal—such as a hacksaw blade—held above the defective coil, so will the shorted stator coil hold an attraction for a strip of steel. In making this test the steel strip must be held to one side of the growler so that it will cover one side of the coil while the growler is covering the other side. Thus

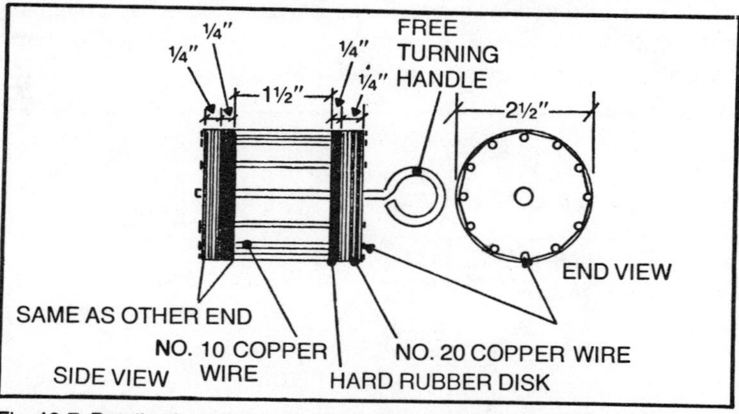


Fig. 13-7. Details of a test rotor for determining the direction of rotation of a rotor in advance of assembly.

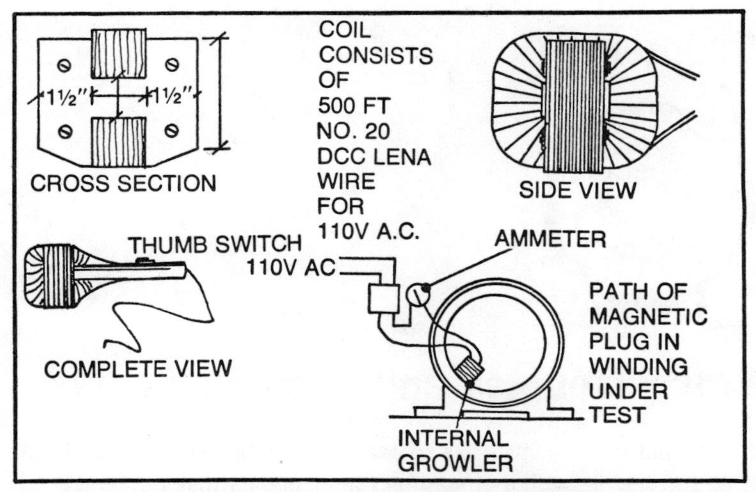


Fig. 13-8. Details of an internal growler for locating shorts and grounds in stators.

the internal growler and the steel strip must be over corresponding sides of the same coil while making the test, and both must be moved around while maintaining their relative positions.

The laminated core of the internal growler can be made from a section of the core from a small direct-current armature by cutting off some of the legs, or it can be shaped from the laminations of some power transformer, such as it is used in radio work. Flexible leads should be provided, and it is well to have an easy-to-reach on and off switch fitted to the handle of the tester. Grounded windings can also be tested with this device by proceeding as follows:

To test for grounds in a stator winding, place the internal growler inside the opening and touch each stator lead to the stator frame. If a spark occurs, that circuit is grounded. If the stator coils are not as yet connected together, or where there is more than one winding, each coil lead or winding lead must be tested separately.

If a ground is discovered in one or more of the stator windings, the exact slot in which the ground is located can be found in the following manner:

Ground one end of the defective coil or winding securely to the stator frame. Next make the growler and hacksaw blade test around the inside of the stator. The slot over which the hacksaw blade vibrates is the one in which the ground is located.

An internal growler is a great time and trouble saver for the motor shop. Like many other "gadgets," it is well worth the labor and thought necessary to construct it.

# Chapter 14

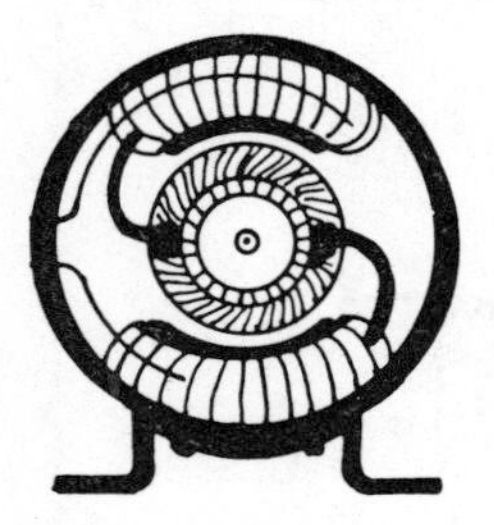

# Brush Troubles and Their Remedies

Every motor repairman has to contend with brush troubles from time to time, as well as the numerous troubles that can be traced directly to improper brush functioning. Altogether too many people have the idea that a brush is just a brush, and that all that is necessary is to roughly fit any handy piece of carbon and insert it in the brush holder. The result is that a large number of motors, especially of the smaller sizes, are continually coming into the shop with commutators burned, pitted, grooved and even worn completely through.

The material from which brushes are made can roughly be classified as follows: Pure carbon brushes, carbon-graphite brushes, graphite brushes, electrographite brushes, and brushes made from metalgraphite composition. The physical properties of these materials or compositions must be taken into account when selecting brushes for a given job.

The resistance of the brush material is a very important factor in many applications. The specific resistance of brush material is its resistance in ohms per cubic inch. In the laboratory this is measured by taking a brush one inch square and taking delicate measurements from opposite sides. If the resistance of the material is too great for the work the brush must do, the brush will heat up in service.

The second important factor in brush design is current carrying capacity. A brush should be able to carry its rated ampere load without causing the operating temperature to rise more than 50° centigrade under usual conditions. The carrying capacity of a brush is determined by both the specific resistance of the material from which it is made, and by the area in contact with the commutator and holder.

The peripheral speed of the commutator also has a decided bearing on brush selection. The higher the peripheral speed the greater the need for higher carrying capacities, and also, the need for better lubrication of the brush. Carbon or carbon-graphite brushes are best suited to slow or medium speed equipment where the carrying capacities are not too high. Electro-graphite brushes are adapted to high peripheral speeds and where commutating characteristics must be of the best. Electro-graphitic brushes are nonabrasive and are best fitted for use on undercut commutators, or where commutator slot insulation does not require an abrasive brush.

Graphite brushes are very soft and meet the requirement of very high operating speeds as well as where the current carrying capacities must be unusually high. As a rule, commutators must be undercut for graphite brushes, for the reason that the brush material is too soft to keep the slot insulation worn down even with the commutating surface. Metal-graphite brushes find general use in low voltage direct-current machines, such as starting motors of various types, and are also used on alternating-current slip ring motors and generators.

## ABRASIVENESS

The abrasiveness of a brush does not necessarily depend on the hardness of its material. A soft grade of carbon brush may be very abrasive, while a hard brush of another type will cause little commutator wear. A certain amount of abrasiveness in a brush is required to keep both the surface of the brush face and of the commutator clean and polished. The abrasiveness of any brush is also influenced by the peripheral speed of the commutator, and by the pressure applied by the brush spring.

The contact drop, or loss of voltage between the face of the brush and the surface of the commutator, depends on the brush material, speed and pressure. The contact drop of almost any brush can be decreased by increasing the spring pressure, but as this is done brush friction is increased. An increase of brush friction may cause serious heating and undue commutator wear. Where the contact drop of a brush is excessive it would be better to change to a less resistant type of brush, than to exert more pressure.

Brush pressure is usually measured in terms of pounds per square inch of brush contacting surface, and as stated before, should be adjusted to a value where a compromise is effected between excessive contact drop and excessive brush friction. A safe rule to follow is to use the lowest brush pressure possible without sparking. This will very seldom be less than 1¾ pounds per square inch of brush contact surface except with very soft graphite brushes, where it may be slightly less.

The motor repairman is seldom able to accurately determine the hardness, abrasiveness, resistance or contact drop of brush material, but he is able to measure the brush pressure on the commutator. This is easily done with a brush spring tension scale, as shown in Fig. 14-1. The hook of the scale is slipped under the outer end of the brush spring and the scale is pulled away from the brush at an angle that would cut through the center of the commutator. The brush arm or spring should barely be raised from the brush when the reading is taken direct in ounces and pounds from the body of the scale. The exact area of the brush surface must be calculated in square inches.

Brushes form such an important part in motor repair service work that every motor shop should maintain a reasonable stock of brushes designed for the more popular types of motors. These brushes, especially for the fractional horsepower motors, are not expensive. Because of the many physical properties of different brush materials, and because the nature of the service differs in different types of motors, it goes without saying that the dealer's brush stock should come from the motor manufacturer or from one of the several reputable factories specializing in the making of well engineered brushes for replacement purposes.

It is impossible for even the large motor shop to stock brushes for every make and model of motor that may come in for service. The time element is often of great importance and it is not always possible to order a new set of brushes and get them in time to suit the owner of the motor. For this reason every motor repair shop should also stock a few sizes of sheet brush carbon.

## MAKING BRUSHES

Carbon sheets of a universal or general purpose grade are available, and usually come in a 4″ × 6″ or 4″ × 8″ size, and in an assortment of thicknesses from ¼ to 1″ or more. Round or square carbon rods of all diameters in one foot lengths can also be bought, and solve the problem of making odd brushes for very small motors, such as are used on fans, drink mixers, etc. With this material at hand any sort of brush can be cut and shaped and, while it may not be the exact type of brush needed, at least a set can be made to "pinch hit" until the proper grade can be obtained.

With a little work the motor repairman can make up brushes with shunts—often improperly called "pig-tails"—where they are required. The duty of a shunt, of course, is to carry the current directly from the proper motor lead to the brush, or vice versa as the case may be. In this way the brush holder, the brush spring or the brush spring arm is relieved as an electrical conductor, and a much better circuit is assured. Shunts are attached to brushes in four ways: They are molded in place when the brush is made; they are fastened in a hole drilled in the brush by means of a screw or pin; they are bolted in place, or they are cemented into a hole drilled in a brush. See Fig. 14-2.

The first way is out of the question for the average repair shop, and the second method is delicate and subject to considerable breakage. It is easy enough to bolt shunts to brushes, but except in the larger sizes there is no room for this type of connection either on the brush or in the confines of the brush holder. Therefore the most practical method for the average shop to follow in fitting shunts to shop made brushes is to cement them in place. Both shunts and shunt cement can be purchased from several of the large brush manufacturing concerns.

In addition to shunts, many of the brushes used on large motors are equipped with hammer or lifting clips. The hammer plate on a brush is a means of preventing fracture of the brush material, or excessive hollowing due to the action of the brush spring arm, grinding on the top of the brush. Lifting clips, as the name implies, are used in applications where the brushes are automatically lifted from the commutator at certain times. Some brushes are fitted with

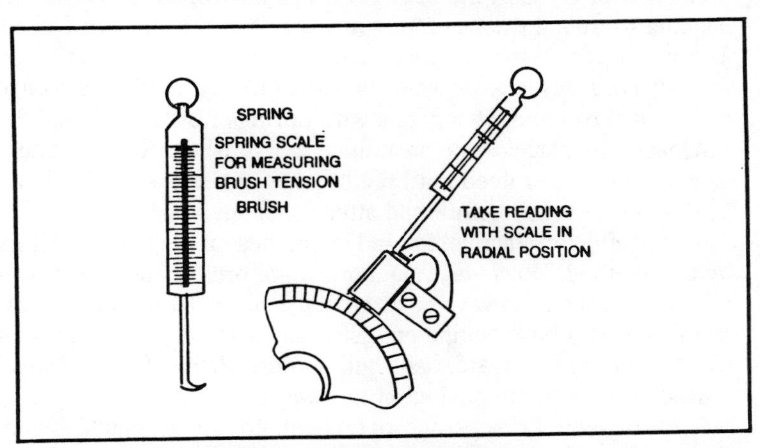


Fig. 14-1. Measuring brush spring pressure or tension.

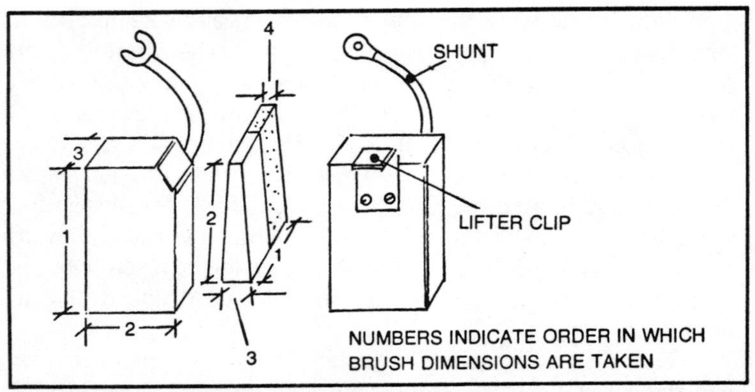


Fig. 14-2. Illustrating brush parts and standard method of measuring square and wedge shaped brushes.

a shunt and a combination hammer and lifting plate. When the need for a set of brushes of this type is imperative, the hammer or lifting clips from the old brushes can be salvaged and reriveted to the new brushes.

## SANDING BRUSHES

Selecting the right type of brush is a very important factor in good performance, but no more so than proper installation and adjustment. This applies with equal force to small as well as large motors. To prevent sparking, excessive wear and chattering, every brush should be sanded to fit the curvature of the commutator. This is easily done on many motors by cutting a strip of 00 sandpaper and banding it around the commutator before the brushes are placed. The sandpaper must be wound around the commutator in a direction that will cause it to tighten when the armature is turned in its normal direction of rotation. String, fine wire or rubber bands will hold the sandpaper in place while installing the brushes. The armature should then be revolved until the brushes are well seated. Figure 14-3 shows a brush before and after sanding.

It is often impractical to sand in brushes on very small motors by this method. However, round or square brushes used in miniature motors should always be fitted to the curve of the commutator, as this cheaply built equipment can stand less arcing as a general rule than can more substantially built motors. Proper fitting of small brushes can be accomplished in this way.

Install new brushes and operate motor for a minute or so. Remove brushes and inspect the contact surface. The glazed part indicates the section of the surface that is rubbing on the commutator. Wrap a piece of fine sandpaper around an object approximately the same diameter as the motor commutator. Rub the end of the brush on this so as to grind away the glazed portion. Reinsert brush in motor and run motor for another minute, then inspect. Keep grinding away glazed portions of the brush until a test shows that at least 75% of the contact surface is bearing on the commutator.

On vertical commutators the brushes may be sanded with a flat strip of fine sandpaper drawn back and forth under the brush. Where the brush holder is attached to the armature shaft and is free to rotate the brushes can all be sanded at once if a circular disk is placed against the commutator face before the brush holder and brushes are installed. After the sanding operation the paper can be torn out. This, of course, is done with the armature out of the machine. Where the sanding of the brushes is done inside the assembled motor or generator, care must be used to blow out all the accumulated carbon dust from the operation.

Motors or generators having adjustable brush holders, and where brush troubles are experienced, should be checked for the angle at which the brushes are set. The proper setting depends to a great extent upon whether the armature rotation is against the heel or the toe of the brush. If the brushes are set for a leading position, the angle, as a general rule, should be in the immediate neighborhood of 35°. If the brushes are in a trailing position the angle may vary from 10 to 25°. Trailing, leading and radial positions of a brush are shown in Fig. 14-4.

Brush spacing around the commutator is a source of trouble on certain critical motors and generators. Incorrect spacing is usually

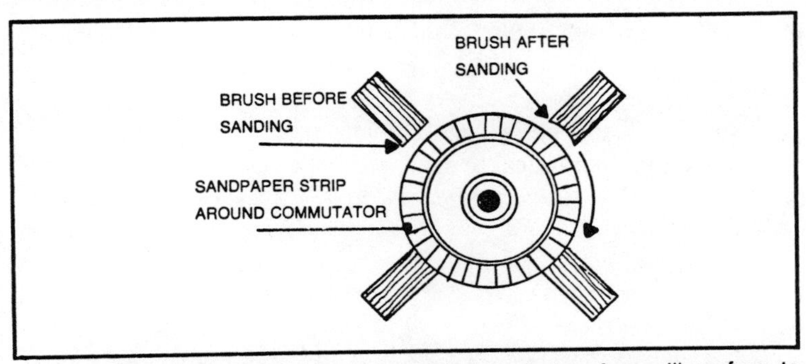


Fig. 14-3. Method of sanding brushes so that contact surface will conform to curvature of commutator.

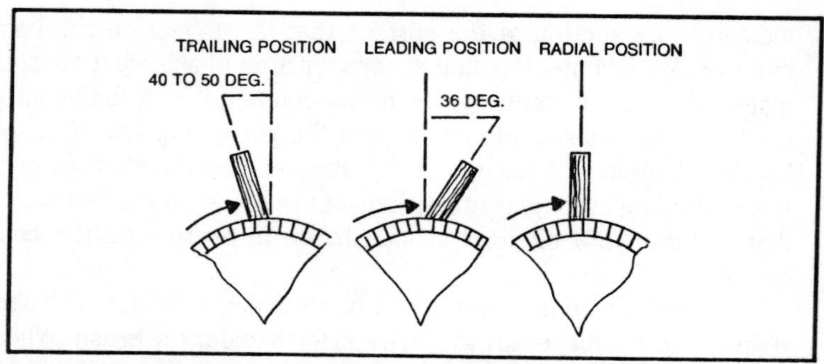


Fig. 14-4. Brush angle and setting on commutator.

caused by careless assembly after a repair job, and shows up worst on interpole machines. An easy way to check for correct brush setting is to wind a strip of paper around the commutator and then mark off the position of each brush. When the paper is removed the distance between brushes is easily measured.

Unusual sparking of the brushes is always an indication of trouble and can most often be traced to one or more of the following causes: High mica between commutator bars; weak spring pressure; incorrect spacing of brushes; oil and/or dirt on commutator; flat spots or commutator out of round; wrong type of brushes; brushes off the neutral; defective field or armature coils; brushes spanning too many bars; brushes tight or gummed in holders; brushes having too low a contact drop; brushes chattering; overload of machine; worn armature bearings; unequal air gaps at poles; vibration of the motor; loose or out of line brush holder studs; loose connections inside motor or at brushes.

# Chapter 15

# Insulating Varnishes

As rewinders and electrical repair men, we are greatly concerned with the subject of insulation. An insulator is basically a material whose electrons are not easily moved about, hence no electric current can easily flow through an insulator. Without insulation in some form or other we would have very few pieces of electrical equipment that would operate. Air is an insulator. So are glass, porcelain, mica, plastic, and rubber. The degree to which a substance is an insulator depends on how much potential it can withstand without breaking down. When the insulation on a wire, for example, is strained to its limit with a high voltage, the insulation degenerates and arcing occurs. To further increase the insulating ability of enamel used on magnet wire, other forms of insulation are applied over finished windings such as tape and varnish.

Varnishes are not only good insulators themselves, but they serve to protect other forms of insulation and to keep their insulating properties functioning. A varnish is essentially a liquid coating which is applied to a surface, which, when dry, forms a hard film. If the average electric motor were not protected by a penetrating coat of good insulating varnish, the action of vibration, heat, water, oil, dirt, and acid fumes, any one or all of them, would soon cause a complete failure in the field or rotor circuits. Some motors designed for use in especially harsh environments depend on a special varnish to make the motor last, where an ordinary varnish in the same application would wear out in just a short time.

If a motor is wound with paper slot wedges, the insulating varnish serves yet another purpose. Plain paper would become brittle and flaky or wet and soggy if it was not sealed in varnish. Plastic strips, which are used on most newer motors, are not susceptible to this, but the varnish cements them in place to prevent vibration from loosening the winding.

Because insulating varnishes are used in relatively small quantities as compared with other motor rewinding materials, and because a knowledge of these varnishes involves principles of chemistry rather than mechanics, many rewinders do not know why one type of varnish is good for one purpose and bad for another. Because of this, a description of various varnish types is in order.

It might be said loosely that a varnish is a paint without a pigment. A little more descriptive statement would be that a varnish is a mixture of gums and resins dissolved in a solvent. There are four main constituents of all varnishes: a thinner or solvent, a resin or adhesive, a drier, and various other forms of oils. There are three main types of varnishes - oil or oleoresinous varnishes, spirit varnishes, and water varnishes.

The water varnishes are almost useless for electric motors since they can readily be dissolved by moisture. A winding coated with a water varnish would not be able to resist water or oils to any great extent, and eventually the winding would fail.

## SPIRIT VARNISHES

Spirit varnishes are quite oil-proof, on the other hand. The most common form of spirit varnish is shellac (the name shellac and other such names refer to the resins used in the varnish). Shellac and other similar resins are dissolved in a thinner, usually alcohol, these forming the group known as spirit varnishes. The solvent, when the varnish is applied, evaporates in a few hours and leaves the gum base behind. The adhesive and insulating qualities of the base depend on exactly what type of resin is used.

The film left by the spirit varnishes is quite brittle. Just coat a piece of sheet metal with shellac, let it dry, bend it, and see what happens. The coating cracks and peels. The same thing happens on a motor winding. Being nothing but a gum, the solvent, and a little oil, spirit varnishes have little of the toughness and elasticity of a good insulating film. When the varnish begins to crack, moisture and other corrosive elements can penetrate the surface to begin working away at the windings. The cracking can also pull the enamel off the wire, thus leaving bare conductors exposed.

## OLEORESINOUS VARNISHES

The most successful varnishes for the impregnating of windings are the oil type varnishes. This group, technically known as oleoresinous varnishes, are composed of a gum or resin, drying oils, and a thinner. In the manufacture of these varnishes, the resin and the oil are given a special heat treatment before the thinner is added. The oil contributes to the flexibility of the base. It may be something such as linseed oil or wood oil. Originally, the resins used were natural gums, but these have poor qualities and often showed poor consistency from one mixture to another. They would swell and stretch and distort the finish. Newer types of oil varnishes use synthetic resins such as epoxy, acrylic, alkyd, and silicone. These resins are what gives each type of oil varnish its name.

The drying agents used are usually salts of some form, and the thinner is often a refined spirit or simple kerosene. The purpose of the thinner is to spread the gums into a thin film for penetration into all parts of the winding. The actual drying of an oil varnish is a two step process. First, the solvent evaporates, leaving the heat treated gums and oils in a thin film on the surface of the work. The remaining gums and oil still in a moist state, then undergo chemical changes which in a certain period of time will cause them to solidify and harden. The application of heat at this point produces a much more rapid action, and, in most varnishes, produces a more desirable and durable film.

The ratio of gum to oil will determine how fast or how slow a certain varnish dries, and also how flexible it is. If the proportion of resin to oil is large, the varnish will dry quickly, but will be brittle. If the amount of resin is small compared to the amount of oil, the varnish will dry slowly but remain more flexible.

While a varnish can be air dried, baking speeds the operation. Baked types of varnishes have general use in the manufacture and repair of electrical machinery. The oil varnishes are more elastic, water resistant, and durable than the spirit type. They are used almost exclusively in the field of motor rewinding.

The oil type varnishes are made in four general types known as clear air drying, black air drying, clear baking, and black baking. Clear colored resins are used for the clear types, while asphaltic materials are used for the black varnishes. The terms "clear" and "black" do not refer to colors, but to transparency. Clear varnishes may have various tints in them depending on the resin used, while the black varnishes may be some other color than dark black.

No one type of varnish is universal, that is, none is the best for every purpose of flexibility, temperature extremes, oil resistance, etc. For the small shop, where the volume of work is limited, it is impossible, of course, to carry a special varnish for each type of work, so a compromise must be made by using perhaps two of the general purpose varnishes.

If the motor is of the newer, high temperature type, it must be coated with a varnish which will be able to withstand the heat without damage. Such a varnish requires a long baking period and will form a hard film. Windings that are subject to severe vibration will hold up better with a varnish that is extremely flexible. This type of varnish will have a short baking period. A varnish of this type, however, will not have the maximum in oil proofing qualities. When a motor has to be repaired in a hurry, an air drying varnish is mostly used regardless to the type of service because of its quick drying time. Emergency repairs place less importance on the insulation.

Usually, the hard drying varnishes are more oil proof than the soft drying type. Baking and air drying varnishes differ in percentage of driers used, and also in the oil resin ratio. Air drying varnishes carry a smaller amount of driers than the baked type.

Alkyd resin varnishes are extremely popular and serve for general purpose use. The epoxy resins are somewhat stronger and are used ın larger motors. Sometimes, plain epoxy adhesive is smeared around the inside of the windings after the varnish is applied to give added strength.

## APPLYING VARNISHES

Just as important as selecting the right type of varnish is applying it in the right way. Ordinary varnish right from the can is often too thick to flow in to all the cracks and crevices of the winding. The use of thinner, therefore, is sometimes necessary. As already mentioned, varnishes are chemical compounds and can be made unstable under adverse conditions. Temperature extremes (freezing or boiling), foreign matter, or adding the wrong kind of thinner to a varnish can spoil it and make it unfit for use.

Varnishes, of course, are extremely flammable and the fumes they give off can be dangerous. Because of this, all varnishes must be kept tightly capped and away from open flames. When a dipping tank is used, it should be provided with a tight cover. A central ventilation system in the shop to exhaust all fumes is a good precaution. Care at all times is another.

Varnishes can be applied with a sprayer, a brush, or by dipping. Dipping is the best way providing that the motor is able to be placed in a tank. The big advantage to dipping lies in the fact that the varnish is able to penetrate and saturate all parts of the winding.

Before the stator is dipped, it is preheated in a oven for two to four hours at a temperature of about 100° C. This is to remove every trace of moisture from the winding so that none will be trapped inside, and it brings the core up to a temperature that will thin the varnish slightly so that it will flow freely into all recesses. Since the varnish itself can be damaged by high temperatures, it is important to have enough varnish in the tank to prevent its temperature from rising more than a few degrees when the armature or stator is immersed.

To make a dipping tank, a large wooden chest may be used, or one may be built. The dipping tank does not have to be really heat resistant, but it must be leak proof. A sheet metal liner is placed on the inside, as shown in Fig. 15-1. A valve should be fitted to the base, and a tight fitting cover should be used for the top. Varnishes can be reused so long as they do not become too old or dirty. A drain grid can also be placed in the tank, allowing motors to drip off excess varnish when they are pulled out. The grid can be a permanent feature, or it can be placed in the tank when needed, so that it will not take up space from the dipping portion itself.

There is one disadvantage to the tank shown, and that is the large airspace above the surface of the varnish. Normally, a varnish lasts longest with minimal air exposure, so some form of floating lid can be designed to prevent the varnish at the top from becoming gummy. A large piece of sheet plastic can be cut to fit inside the tank. Before closing the top, the floating lid is placed on the varnish where it reduces the surface area exposed to the air inside the tank. An alternative is to drain the tank after each use, but this will waste

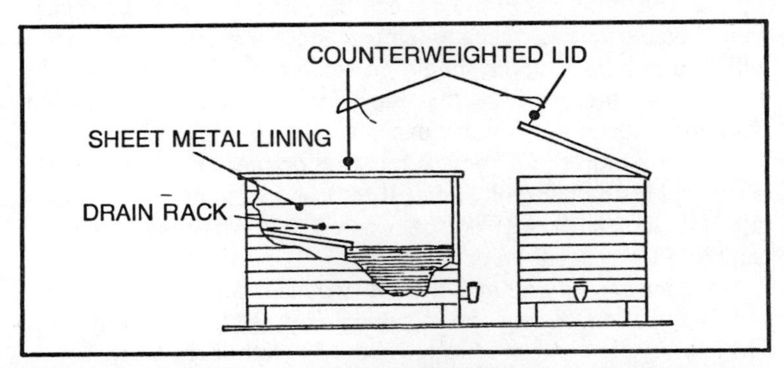


Fig. 15-1. This dipping tank with draining facilities will serve the small motor shop adequately. The lid should fit tightly and a valve should be provided for drawing off the liquid for cleaning. The size of the tank will be determined, of course, by the size of the work handled.

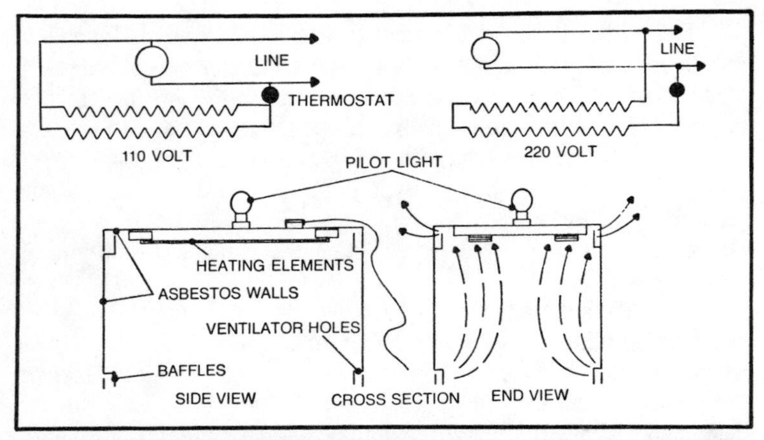


**Fig. 15-2. This portable hood type baking oven is convenient for small work. The sides and top are constructed of stiff asbestos board held in a light angle iron frame. The enclosed type heating units are controlled by a thermostat. The oven is placed over the work to be baked.**

varnish by forming films which will build up on the storage cans and the drain valve, as well as inside the tank.

To perform the dipping operation, the hot frame is lowered into the tank and allowed to sit there while the varnish penetrates all the places inside the winding. The best rule is to leave the work in the tank until all bubbling ceases, but a much longer period will do no harm. If the varnish is to be brushed on, set the motor so that one end of the stator faces upwards and flow the varnish on heavily, letting it run down into the slots. Then turn the motor over to expose the other side and repeat the process. This procedure should be performed over a catch pan, since a good deal of varnish will drip out. The motor should be rotated several times and re-soaked with the brush so that the winding is fully impregnated. Spraying is done in a similar manner.

After a motor is dipped, it must be drained. Putting it on the drain grid inside the tank allows it to drop its excess varnish back into the pool so it will not be wasted. The draining should be complete for reasons of economy.

Baking can be accomplished in two main ways, by convection and by radiation. A convection baking oven is shown in Fig. 15-2. The heating elements are located above the motor and warm the air inside, which in turn dries the windings. There is a certain amount of infrared radiation which strikes the motor directly from above, and thus heats the motor frame by conduction as well. When placed in an oven, the coils should be placed upside down from the position in which they were drained, so that the heat will cause the varnish to flow back over the motor to produce a more even film. In this way, thick globules will be eliminated.

The oven should supply an even heat to the motor. There should be plenty of ventilation that will carry away the solvent fumes and allow fresh air to enter the oven. A forced air system is the best. A thermostat control could be used to keep the temperature at approximately 100° C. The time required for the baking will vary with the varnish, the type of work, and the thickness of the film. Thin films are easier to dry and oxidize than heavy films. Different film thicknesses are required for each motor, and this in turn will be determined by how much the varnish is thinned. Ordinarily, one coat of varnish is insufficient, and so the process is repeated. This may be from one to several more times, depending on the severity of the operating conditions.

Certain requirements, such as anti-fungal treatments, are sometimes added to varnishes to make them more rugged. The type of resin used also has a bearing on what kind of resistance it will have to acids, corrosive atmospheres, etc. Not only does the varnish protect the winding, but it also protects the inside of the frame as well, so it is important to keep scratches out of finished motors.

An infrared heating lamp can be used to dry a motor winding, as shown in Fig. 15-3. Normally, several lamps are placed around the motor facing inward at the core. Again, it is important to carry away fumes with fresh air, because a solvent trapped in its own vapors will not dry properly. A caution about the lamps is that they must be

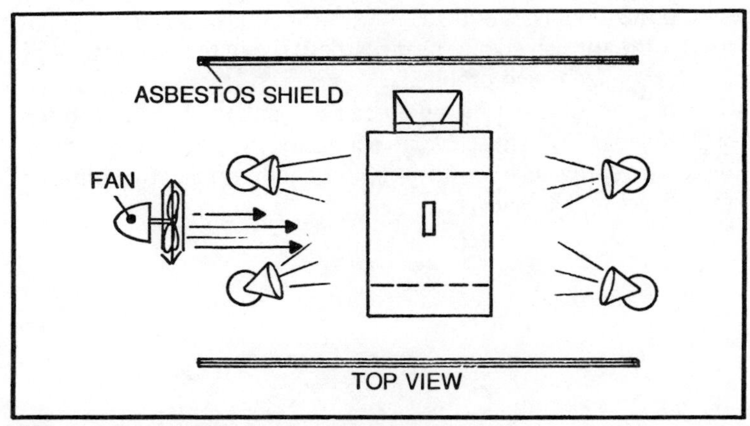


Fig. 15-3. Placing of heat lamps to dry a winding.

kept reasonably far away from the coils so as not to actually burn the varnish or the windings. When the last bake is complete, the winding is allowed to cool for as long as it is practical. The varnishes may still be slightly soft until they cool, so it is important to let them set undisturbed until they become rock hard.

Most varnishes are too thick for general use and must be thinned to reduce their specific gravity. The amount that a varnish must be thinned can be found by experimenting, or it can be done using a hydrometer that is calibrated for measuring liquids that are *lighter* than water. Varnishes will thicken when they are cold and soften when they are heated, so it is important to do all the thinning and measuring at one particular temperature, say 70° F. If the temperature is different, a correction factor must be added or subtracted from the hydrometer reading. To do this, add .0004 to the reading for each degree above the 70 mark, and subtract it for each degree below.

Alcohol is the thinner used for spirit varnishes. A good solvent for many of the commercial baking varnishes is procurable from many of the larger oil companies under the name of "solvent" or "light naphtha." The distillation end point should be below 160°F., and the density from 54° to 58° Baume. If this is not available, the recommended thinner specified on the can can be used. Kerosene is also a popular thinner. Turpentine cannot be used because it is an enamel solvent and will soften the coatings on the wire. When mixing, both the thinner and the varnish should be of the same temperature. The thinner should be added slowly and mixed thoroughly. Careful checks will prevent over-thinning. If the varnish is old and oxidized, or if one of the two is cold, curdling will result. If the mixture is allowed to warm up, this may cease, but if it persists, the varnish should be discarded. Using the wrong type of thinner will also cause problems.

A high voltage acrylic spray can be applied if desired to the completed winding. These can be purchased in radio supply stores. A dielectric spray such as this provides another insulating film and prevents arcing and shorts.

# Index